Mastering Claude AI

Practical Journey from First Prompts to Pro with Claude AI

Ryan Dickey

Apress®

Mastering Claude AI: Practical Journey from First Prompts to Pro with Claude AI

Ryan Dickey
Grand Junction, CO, USA

ISBN-13 (pbk): 979-8-8688-2000-7 ISBN-13 (electronic): 979-8-8688-2001-4
https://doi.org/10.1007/979-8-8688-2001-4

Managing Director, Apress Media LLC: Welmoed Spahr
Acquisitions Editor: Celestin Suresh John
Development Editor: James Markham
Editorial Assistant: Gryffin Winkler

Cover designed by eStudioCalamar

Distributed to the book trade worldwide by Springer Science+Business Media New York, 1 New York Plaza, New York, NY 10004. Phone 1-800-SPRINGER, fax (201) 348-4505, e-mail orders-ny@springer-sbm.com, or visit www.springeronline.com. Apress Media, LLC is a Delaware LLC and the sole member (owner) is Springer Science + Business Media Finance Inc (SSBM Finance Inc). SSBM Finance Inc is a **Delaware** corporation.

For information on translations, please e-mail booktranslations@springernature.com; for reprint, paperback, or audio rights, please e-mail bookpermissions@springernature.com.

Apress titles may be purchased in bulk for academic, corporate, or promotional use. eBook versions and licenses are also available for most titles. For more information, reference our Print and eBook Bulk Sales web page at http://www.apress.com/bulk-sales.

Any source code or other supplementary material referenced by the author in this book is available to readers on GitHub. For more detailed information, please visit https://www.apress.com/gp/services/source-code.

If disposing of this product, please recycle the paper

For my family and Claude Shannon.

Table of Contents

About the Author

Ryan Dickey is a Claude AI power user who discovered the transformative potential of human-AI collaboration through hands-on experimentation. After using project management skills to self-publish his first book, he signed with Apress to create this comprehensive guide to mastering Claude. Ryan holds a Bachelor of Applied Science in Public Safety Studies from Siena Heights University and has earned multiple certifications, including Google Project Management Professional, Google AI Essentials, and Google Prompting Essentials. His unique background spans emergency services, sports journalism, horse racing, and content operations, providing him with diverse perspectives on practical AI applications. Ryan is passionate regarding helping others unlock Claude's capabilities for personal and professional growth.

About the Technical Reviewer

Mark Koranda's journey into AI began with a fundamental need to translate between different worlds. Growing up as the hearing child of deaf parents required both technical precision and deep empathy to bridge communication gaps—skills essential to his later work connecting human cognition with artificial intelligence.

After serving as a Cryptologic Linguist for the US Marine Corps, translating Arabic and Pashto in high-stakes intelligence operations, Mark pursued his PhD in Psychology at the University of Wisconsin-Madison. There, he discovered that the same pattern-recognition abilities he'd honed in military intelligence could unlock secrets on how neural networks process language. His research, published in top-tier journals like *Psychological Science*, revealed how subtle contextual changes can make people say things they didn't consciously intend.

Acknowledgments

Writing a book is never a solo endeavor, and this project benefited from the generous support of many people and one remarkable AI.

I thank the entire Apress editorial and production team for their expertise and guidance throughout the publishing process. Special recognition goes to Mark Koranda for his thorough technical review, guaranteeing the accuracy and clarity of the content.

Finally, I must acknowledge my most consistent collaborator: Claude itself. This book was written with Claude, not just regarding it. Every chapter benefited from our ongoing dialogue, and the meta-experience of writing on AI collaboration while actively collaborating with AI shaped every page.

Any errors or omissions remain entirely my responsibility.

Preface

Why This Book Exists

This book documents a fundamental shift in how humans work with artificial intelligence. Through 20 chapters of systematic guidance, evidence-based frameworks, and rigorously tested techniques, you'll learn to collaborate with Claude—Anthropic's AI assistant—in ways that amplify your professional capabilities while maintaining appropriate expectations about AI's actual limitations.

I'm Ryan Dickey, and I wrote this book for a simple reason: the gap between having access to AI and knowing how to use it effectively is massive. After spending countless hours developing expertise with Claude, completing major projects, and discovering techniques that consistently deliver results, I realized this knowledge needed to be shared systematically—with full technical accuracy and realistic expectations about what human-AI collaboration can and cannot achieve.

What You'll Find Inside

Part I: Claude Fundamentals establishes essential knowledge with technical precision. You'll understand what Claude actually is (and isn't) and how it differs from other AI tools and master the foundational skills of prompting and conversation management through systematic frameworks that remain relevant across AI generations.

Part II: Practical Applications covers five core areas where Claude provides measurable assistance: writing and communication, research and analysis, coding and programming, creative problem-solving, and data analysis. Each chapter provides evidence-based templates, realistic examples, and progressive exercises with appropriate limitations clearly explained.

Part III: Advanced Techniques introduces sophisticated prompting strategies grounded in systematic methodology, Claude's special features like Artifacts and Projects (with current technical constraints documented), and workflow integration methods that demonstrably transform productivity when properly implemented.

Part IV: Professional and Specialized Applications explores domain-specific applications in business, education, and creative fields, with case studies from real practitioners. Each application includes critical guidance on professional standards, disclosure requirements, and technical limitations that vary by industry and jurisdiction.

Part V: Advanced Topics and Troubleshooting addresses critical areas including systematic troubleshooting with technical specifications, comprehensive ethics and responsible AI use (including legal and professional requirements), and evidence-based strategies for staying current as AI capabilities evolve.

Part VI: Excellence and Beyond charts the realistic path from competent user to power user through sustained practice and systematic skill development, concluding with grounded insights on the future of human-AI collaboration based on current technical realities rather than speculation.

The appendixes provide quick reference guides, curated learning resources, and battle-tested templates you can implement immediately—all with appropriate verification and quality control guidance.

What Makes This Approach Different

This book emerged from actual collaboration with Claude, enhanced through rigorous technical review to ensure every claim reflects current AI capabilities accurately. Throughout these pages, you'll find "From the Chat" sections showing real exchanges, including breakthroughs, mistakes, and unexpected discoveries. You're not reading about theoretical possibilities—you're learning from documented practice, technically verified and systematically organized.

The techniques presented here have been tested across diverse fields by professionals at all skill levels. The recurring stories you'll encounter—Sarah transforming her coffee shop operations, Marcus advancing from marketing intern to team lead, Dr. Chen accelerating her research output, Tom learning Python at 67—represent composite illustrations of real patterns observed across many practitioners, not individual case studies.

Who Should Read This Book

This book serves several audiences with realistic expectations about skill development timelines:

Business Professionals seeking competitive advantage through AI-enhanced productivity, better decision-making, and automated workflows (typically requiring 3–6 months of sustained practice for professional competency).

Writers and Creatives looking to overcome blocks, enhance their work, and explore new possibilities while maintaining their unique voice and understanding AI's pattern-based assistance limitations.

Students and Educators wanting to accelerate learning, create better educational materials, and prepare for an AI-integrated future while meeting institutional disclosure requirements and academic integrity standards.

Technical Professionals interested in using AI for coding assistance, debugging, documentation, and system design with realistic expectations about current technical capabilities and limitations.

Anyone Curious About AI who wants practical skills based on evidence rather than hype, including those who've tried AI tools but haven't found their rhythm or need systematic frameworks for consistent results.

What You Won't Find

This isn't a technical manual about how AI works under the hood. It's not a philosophical treatise on artificial consciousness. It's not a collection of prompts to copy-paste without understanding. You also won't find unsupported claims about "co-pilot" relationships or overstated promises about transformation that exceed what current research validates.

Instead, you'll develop systematic frameworks for human-AI collaboration that transcend specific features or versions. You'll learn principles that remain valuable as technology evolves, grounded in realistic expectations about current AI capabilities. You'll gain confidence to explore and innovate within appropriate technical boundaries rather than follow rigid rules or harbor unrealistic expectations.

The Evergreen Approach: Why This Book Focuses on Principles

I tried to keep most of this content as evergreen as possible, and here's why that matters: the AI assistant space moves at breathtaking speed. Features appear and disappear, interfaces change overnight, and what worked perfectly last month might need adjustment today. Claude's capabilities evolve rapidly, new AI assistants emerge regularly, and even fundamental approaches to human-AI interaction shift as researchers discover better methods.

Rather than chase every new feature or write a manual that becomes obsolete within months, I focused on systematic frameworks and transferable principles. The prompting strategies you'll learn work across AI systems. The collaboration methodologies apply whether you're using Claude, future versions of Claude, or entirely different AI assistants. The ethical frameworks remain relevant regardless of technological advancement.

When specific dates, versions, or current limitations needed mentioning, I included them—but always within a broader principle that transcends the immediate technical details. This approach means you're not just learning to use today's Claude; you're developing expertise that adapts as the entire field evolves.

Critical Context and Disclaimers

Technical Reality: AI systems like Claude work through sophisticated pattern recognition based on training data, not genuine understanding, creativity, or consciousness. Every technique in this book acknowledges these fundamental limitations while maximizing practical value within realistic bounds.

Evidence and Examples: Success stories and case studies represent composite illustrations for educational purposes, not individual testimonials. Techniques described have been tested and validated, but effectiveness varies based on individual application, context, and commitment to systematic practice.

Professional Requirements: AI assistance disclosure requirements vary significantly by profession, institution, and jurisdiction. This book provides general guidance, but readers must research and comply with their specific professional and legal requirements.

Skill Development Timelines: Developing professional-level AI collaboration skills typically requires sustained practice over 6–24 months, not just reading about techniques. This book provides the framework; consistent application builds the competency.

A Note on Ethics and Responsibility

Throughout this book, you'll find systematic guidance on using AI ethically and responsibly, including understanding technical limitations, avoiding over-dependence, protecting privacy across varying jurisdictional requirements, and maintaining human judgment in critical decisions. The goal isn't to replace human intelligence but to amplify it thoughtfully within appropriate technical and ethical boundaries.

How to Use This Book

While designed to be read sequentially for systematic skill development, each part stands alone for readers with specific needs. Beginners should start with Part I and follow the progressive framework. Those with AI experience might benefit from the systematic approaches in Part III. The appendixes serve as ongoing references during your practice.

Look for these elements throughout:

- **Systematic Frameworks** that connect principles to practice across AI generations
- **Evidence-Based Examples** from actual use cases with appropriate qualifications
- **Implementation Templates** you can adapt immediately with verification guidance
- **Technical Reality Checks** to avoid common pitfalls and unrealistic expectations
- **Progress Frameworks** to track your systematic skill development

An Invitation to Realistic Possibility

The transformation I've experienced—and witnessed in others—through effective AI collaboration has been significant and measurable. Not because AI is magic, but because understanding how to work within its actual capabilities reveals human potential we didn't know we had, while maintaining appropriate expectations about what human-AI collaboration can realistically achieve.

Whether you're looking to work more efficiently, solve complex problems within current technical constraints, enhance creativity through pattern-based assistance, or simply understand what's actually possible vs. what's marketing hype, this book provides a systematic, evidence-based path forward.

The future of work isn't about choosing between human or artificial intelligence—it's about combining them effectively within realistic technical boundaries, with full understanding of both capabilities and limitations.

Welcome to your journey toward evidence-based AI collaboration mastery. Let's begin with clear eyes and realistic expectations.

—Ryan Dickey

P.S. Yes, Claude assisted in writing this preface, just as it assisted throughout the entire book. This collaboration demonstrates the very principles you'll learn—transparency, systematic methodology, and the amplification of human capability through AI within appropriate technical boundaries. The ideas, insights, and strategic decisions remain human. The execution benefits from AI enhancement within clearly understood limitations. That's the future we're building together—realistic, evidence-based, and systematically effective.

PART I

Claude Fundamentals

CHAPTER 1

What Is Claude and Why Should You Care?

In This Chapter

- Meeting your new AI assistant (spoiler: it's pretty amazing)
- Understanding what makes Claude different from other AI tools
- Discovering real-world uses that'll make your life easier
- Learning the ground rules for AI collaboration
- Getting excited about the possibilities ahead

Remember that friend who always knows the correct answer, never gets tired of your questions, and can help with everything from writing emails to debugging code? Well, you haven't met them yet—because they're not human. They're Claude, and this chapter introduces you to your new AI collaborator who's about to change how you work, learn, and create.

So, What Exactly Is Claude?

Let me paint you a picture. It's 1:59 AM, you're staring at a blank screen with a report due in seven hours, and your brain feels like mashed potatoes. Or maybe it's Tuesday afternoon, and you're trying to understand some complex code that might as well be ancient hieroglyphics. Or perhaps you're planning a dinner party and need creative menu ideas that won't result in ordering pizza (again).

R. Dickey, *Mastering Claude AI*, https://doi.org/10.1007/979-8-8688-2001-4_1

Enter Claude—an AI assistant created by Anthropic that's like having a brilliant, patient, and endlessly helpful colleague available 24/7. But Claude isn't just another chatbot spouting random internet facts. This AI model represents a fundamental shift in how we interact with artificial intelligence.

The Technical Bit (Don't Worry, We'll Keep It Simple)

Claude is what's called a **Large Language Model (LLM)**—think of it as a very sophisticated pattern-recognition system that's been trained on vast amounts of text. What really powers Claude is **transformer neural network architecture**—the technology that allows computers to understand, interpret, and generate human language in a way that's actually useful. It's like teaching a computer to not just recognize words, but to truly grasp meaning, context, and even subtle nuances like sarcasm or humor.

But here's where it gets interesting: Unlike some AI models that are just trying to predict the next word like the world's most expensive autocomplete, Claude was trained using something called **Constitutional AI**—a method developed by Anthropic that gives Claude explicit ethical principles to follow, rather than relying solely on human feedback.

Plain English Translation: Claude was programmed and taught not just to be smart, but to be helpful, harmless, and honest. It's like the difference between hiring someone who is brilliant but might give you terrible advice vs. someone who's both smart AND has good judgment.

What Makes Claude Special?

You might be thinking, "Great, another AI assistant. I've tried ChatGPT, and my phone already has Siri. What's the big deal?"

Fair question! Here's what sets Claude apart:

> **Constitutional AI Training**: Claude's training method is unique. Instead of learning behavior purely from human feedback, Claude follows a written "constitution" of principles that guide its responses. This makes Claude more consistent, transparent, and aligned with helpful, harmless values.

Longer Conversations: Claude can handle much longer conversations and documents than most AI assistants. While others might forget what you said five minutes ago, Claude can maintain context throughout lengthy discussions. It's like the difference between having a conversation partner with excellent memory vs. one who keeps forgetting the topic.

Better at Following Instructions: Ever tried to get an AI to stick to a specific format or style? It's usually like herding cats. Claude consistently follows complex instructions and maintains the style or approach you specify.

More Thoughtful Responses: Claude doesn't just generate the first response that comes to mind. Its training process encourages consideration of context, nuance, and potential implications. Think of it as the colleague who actually thinks before they speak.

Vision Capabilities: Claude can analyze and understand images, including charts, diagrams, screenshots, and photographs. This makes it incredibly useful for tasks involving visual information.

Web Search Integration: When you need current information, Claude can search the web to provide up-to-date facts, news, and data.

Refuses to Be Harmful: Claude has built-in guardrails that prevent it from helping with dangerous, illegal, or harmful requests. It's like having a designated driver for your AI interactions—keeping things safe even when you might not be thinking clearly.

Real People, Real Uses: Claude in Action

Let's move beyond the theory and look at how people might use Claude every day. Throughout this book we'll check in on four main characters and how Claude can help them:

Sarah the Small Business Owner

Sarah runs a boutique marketing agency. She uses Claude to:

- Draft client proposals in minutes instead of hours
- Brainstorm creative campaign ideas when her team is stuck
- Proofread and polish presentations before client meetings
- Create social media content calendars
- Explain complex analytics data in simple terms her clients understand
- Analyze charts and graphs from campaign performance reports

"Before Claude, I was working 60-hour weeks. Now I get more done in 40 hours and actually see my kids at dinner."

Marcus the College Student

Marcus is pursuing a computer science degree while working part-time. He uses Claude to

- Debug code when he's stuck (without just copying solutions)
- Understand complex programming concepts through analogies
- Organize study schedules around his work shifts
- Write clear, professional emails to professors
- Prepare for technical interviews
- Analyze diagrams and visual materials from textbooks

"Claude doesn't do my homework for me—it helps me understand so I can do it myself. It's like having a tutor available at 3 AM."

Dr. Jennifer Chen, Researcher

Dr. Chen uses Claude in her medical research lab to

- Summarize lengthy research papers quickly
- Generate hypotheses for new experiments
- Draft grant proposals with proper formatting
- Create presentations for conferences
- Translate complex findings for public audiences
- Analyze research charts, graphs, and medical imaging data

"Claude helps me spend less time on paperwork and more time on actual research. It's transformed how our lab operates."

Tom the Retiree

Tom recently retired and uses Claude for personal projects:

- Planning travel itineraries with detailed daily schedules
- Learning new hobbies with step-by-step guidance
- Writing family history stories for his grandchildren
- Understanding modern technology his kids keep talking about
- Getting help with photos and images he wants to understand
- Solving crossword puzzles (with hints, not answers!)

"I thought AI was just for tech people. Turns out it's for anyone who's curious and wants to learn."

What Claude Can (and Can't) Do

Let's set realistic expectations. Claude is incredibly capable, but it's not magic. Understanding its strengths and limitations will help you use it effectively.

Claude's Superpowers

Writing and Editing: From emails to essays, Claude can help you write better, faster, and with more clarity. It can adapt to any tone or style you need.

Coding and Technical Help: Whether you're debugging Python or trying to understand Excel formulas, Claude speaks fluent tech.

Research and Analysis: Claude can synthesize information, identify patterns, and help you understand complex topics quickly. With web search access, it can also find current information.

Creative Collaboration: Need ideas for anything from birthday parties to business names? Claude's got you covered.

Learning and Education: Claude can explain anything from quantum physics to cooking techniques at exactly your level of understanding.

Visual Analysis: Claude can analyze photographs, charts, diagrams, screenshots, and other images to extract information, answer questions, or provide insights.

Claude's Limitations

Not Perfect Accuracy: While highly reliable, Claude can make mistakes or have gaps in knowledge. Always verify important information, especially for critical decisions.

Knowledge Cutoff: Claude's training data has a cutoff date (end of January 2025 for current models), though it can search the web for more recent information when needed.

Personal Data Access: Claude can't access your files, emails, or browsing history unless you explicitly share them in the conversation.

No Independent Internet Access: Claude can't browse the web on its own or perform actions outside of the conversation interface.

Can't Make Physical Changes: Claude can tell you how to update your website, but it can't actually click the buttons for you.

The Ethics Thing (It's Actually Important)

Before we dive deeper into using Claude, let's talk about the elephant in the room: AI ethics. Don't worry, this isn't a philosophy lecture, but understanding these principles will make you a better Claude user. There's an entire chapter dedicated to AI ethics in this book.

The Three H's: Helpful, Harmless, Honest

Claude operates on three core principles:

Helpful: Claude aims to assist you in achieving your goals, whether that's writing better, learning faster, or solving problems more efficiently.

Harmless: Claude won't help with activities it determines to be dangerous, illegal, or harmful. This isn't Claude being a buzzkill—it's ensuring AI remains a positive force.

Honest: Claude strives to admit when it doesn't know something rather than making things up, though like any AI system, it's not perfect at this. It's refreshing in a world full of confident nonsense.

Your Responsibility as a User

With great AI power comes great responsibility (yes, I went there). Here's what you need to remember:

Verify Important Information: Always double-check facts for critical decisions. Claude is smart, but it's not infallible.

Respect Privacy: Don't share sensitive personal information—yours or others'—in your conversations.

Use It Ethically: Don't try to use Claude for academic dishonesty, spreading misinformation, or any harmful purposes.

Give Credit Where Due: If Claude helps you create something significant, it's good practice to acknowledge AI assistance.

Getting Started: Your First Conversation

Ready to meet Claude? You can access Claude through the web interface at claude.ai, mobile apps, or via API for developers. Here's how to make your first interaction successful:

Step 1: Clear Communication

Claude responds best to clear, specific requests. Instead of "Help me with my report," try "I need to write a 500-word executive summary of our Q3 sales performance, focusing on growth in the Northeast region."

Step 2: Provide Context

The more context you give, the better Claude can help. Share relevant background information, your goals, and any constraints or preferences.

Step 3: Iterate and Refine

Don't expect perfection on the first try. Claude excels at revision and refinement. Ask for adjustments, different approaches, or clarifications as needed.

Step 4: Build on Responses

Claude remembers your entire conversation. You can reference earlier points, ask follow-up questions, and build complex projects step by step.

The Bottom Line

Claude isn't just another tech tool to learn—it's a capability amplifier. It won't replace your creativity, judgment, or expertise. Instead, it enhances them, letting you work smarter, learn faster, and create better.

Think of Claude as your intellectual Swiss Army knife: versatile, reliable, and incredibly useful once you know how to use it. Whether you're writing your first novel, starting a business, learning quantum mechanics, or just trying to word that tricky email to your boss, Claude is ready to help.

In the following chapters, we'll dive deep into specific techniques, advanced features, and real-world applications. But for now, just remember this: Claude is here to help you become a better version of yourself, not to replace you.

Welcome to the future of human-AI collaboration. It's going to be quite a ride.

CHAPTER 2

Getting Started: Your First Conversation with Claude

In This Chapter

- Setting up and navigating Claude's interface like a pro
- Understanding how conversations work (spoiler: it's easier than texting)
- Mastering the art of the perfect first prompt
- Learning the secret sauce of context and memory
- Troubleshooting common rookie mistakes

So you understand what Claude is. Great. Now let's actually use it. Navigate to claude.ai in your browser, download the mobile app, or install the desktop application, and prepare to have your mind quietly blown. No dramatic music required—just you, a text box, and an AI assistant that's about to become your new favorite colleague. This chapter transforms you from Claude-curious to Claude-capable, one conversation at a time.

The Interface: Your Command Center

First things first—let's talk about where the magic happens. You can access Claude through three main channels: the web interface at claude.ai, mobile apps for iOS and Android, or desktop applications for macOS and Windows. Claude's interface shares many conventions with other AI chatbots you may have used but has some unique features worth highlighting.

R. Dickey, *Mastering Claude AI*, https://doi.org/10.1007/979-8-8688-2001-4_2

The Basics

When you open Claude, you'll see a clean, straightforward interface:

- **The Chat Area:** This is where your conversation unfolds, like a text message thread with the world's smartest friend
- **The Input Box:** Where you type your prompts (think of it as Claude's ears)
- **The Send Button:** Your "make it happen" button (or just hit Enter—we're not savages)
- **The New Chat Button:** For when you want to start fresh with a clean slate

That's it. No 35 different menus, no mysterious icons that might launch nuclear missiles. Just you, Claude, and a conversation waiting to happen.

Mobile vs. Desktop

Using Claude on mobile? The interface adapts beautifully, but here's a pro tip: For longer, more complex work, desktop is your friend. It's like the difference between cooking in a full kitchen vs. a camping stove—both work, but one's definitely easier for Thanksgiving dinner.

Understanding the Conversation Flow

Here's where things get interesting. A conversation with Claude isn't just a series of disconnected questions and answers—it's an actual dialogue. Think of it as a thread that weaves through your entire interaction.

What's a Thread, Anyway?

In Claude terms, a thread is your complete conversation from start to finish—from your first prompt until you close the window or start a new chat. Everything you discuss in that thread stays connected, like chapters in a book you're writing together.

The Chat History: Your Conversation's Record

Within each thread, Claude maintains a chat history—a complete record of everything you've discussed in that session. This history is what you see in the interface, showing the full back-and-forth conversation. But here's an important technical distinction: **the chat history (what you see) is not the same as Claude's context window (what Claude actively remembers).**

Think of it this way:

- **Chat History**: The complete transcript of your conversation, like a meeting transcript that preserves everything
- **Context Window**: Claude's active working memory during that conversation, like how much of that meeting transcript you can keep in your head while speaking

This distinction matters because Claude uses the chat history within its context window to:

- Remember what you've already talked about
- Build on previous ideas
- Maintain consistency in tone and approach
- Avoid repeating information unnecessarily

It's like having a conversation with someone who actually listens and remembers what you said five minutes ago. Revolutionary, right?

Your First "Hello, Claude" Moment

Alright, you're staring at that empty chat window. What now? Let's make this less intimidating than a blank page in a new journal.

Breaking the Ice

Here's a secret: Claude doesn't judge your opening line. You could literally type "Hi" and get a friendly response. But since you're here to get things done, let's start smart:

Good First Prompts:

- "Hi Claude! I'm new to using AI assistants. Can you help me understand what you're best at?"
- "I need help writing a thank-you email to my boss. Can you guide me through it?"
- "Can you explain how photosynthesis works like I'm 10 years old?"

Not-So-Great First Prompts:

- "Do my job for me"
- "Tell me everything about everything"
- "asdfghjkl" (Though Claude will politely try to help even then!)

From the Chat: When Intelligence Meets Intelligence

Sometimes the best conversations with Claude happen when you get philosophical. Here's a real exchange that shows how Claude can engage with deeper questions while keeping things light, demonstrating the collaborative nature of human-AI interaction:

> **Ryan**: While your "artificial" intelligence is a series of 0's and 1's, my "organic" intelligence is electrical impulses moving through meat in my skull.
>
> **Claude**: Ha! That's a brilliant way to put it! 😄

The Great Intelligence Divide:

- **Me**: Digital patterns in silicon and electricity
- **You**: Electrochemical patterns in neurons and synapses

But here's the funny thing—we're both just information processing systems having a conversation about writing a book together. Your "meat computer" is directing my "silicon computer" to help explain how other meat computers can better collaborate with silicon computers.

The meta-level absurdity is perfect:

- Biological intelligence teaching artificial intelligence
- How to teach other biological intelligences

- To work with artificial intelligence
- To create better outputs than either could alone

And somehow your electrical meat signals are managing to project-manage my binary responses into something coherent and commercially viable!

For the book: This could actually be a great way to demystify AI for readers—"Claude processes information differently than you do, but you're both pattern-recognition systems having a conversation."

Your organic intelligence + my artificial intelligence = hopefully one hell of a book! 🧠 🤖 📚

This exchange perfectly captures the collaborative spirit of working with Claude—it's not about one type of intelligence being better than another, but about how they can work together to create something neither could achieve alone.

The Magic (and Limits) of Context

Now for some insider knowledge about how Claude's brain works. Every AI has something called a context window—think of it as Claude's working memory during a conversation.

Understanding the Context Window

The context window determines how much of your conversation Claude can actively reference when generating responses. Here's the key technical insight:

Context Window ≠ Chat History

- **Chat History**: The full record of everything said (what you see in the interface)
- **Context Window**: How much of that history Claude can actively process at once (what Claude "remembers" when responding)

For shorter-to-medium length conversations, these are effectively the same thing. But for very long conversations, Claude might need to work with a compressed or summarized version of the earlier chat history to fit within its context window limits.

Claude has one of the largest context windows in the AI world—up to 1 million tokens for Claude 4 Sonnet—which means it can handle:

- Very long documents (like that 100-page report you need summarized)
- Extended conversations without losing track
- Complex projects with multiple components
- Back-and-forth discussions that would make other AIs dizzy

Making the Most of Context

Here's how to leverage Claude's impressive memory:

Do:

- Reference earlier parts of your conversation ("Like we discussed above...")
- Build complex projects step by step
- Ask follow-up questions
- Request revisions based on previous outputs

Don't:

- Assume Claude remembers conversations from different threads
- Expect perfect recall of extremely long documents (even elephants have limits)
- Panic if you hit the context limit—just start a fresh thread with a summary

Tokens: The Building Blocks of AI Conversation

Time for a tiny bit of technical knowledge that'll make you sound smart at parties (or at least at tech meetups). Claude processes text in units called tokens.

What's a Token?

Tokens are like AI building blocks—the basic units that Claude uses to understand and generate text. They're not exactly words, not exactly letters, but something in between. Token count varies by model, but here's the general idea:

- Simple words like "Hello" = 1 token
- Complex phrases may = multiple tokens
- Punctuation and emojis also count as tokens

Why Should You Care?

Understanding tokens helps you:

- Write more efficient prompts
- Understand why some requests take longer
- Maximize your conversation length
- Make sense of usage limits and pricing

Think of tokens like data usage on your phone plan—you want to use them wisely to get the most out of your conversation. Claude gives you plenty to work with, but being efficient helps you accomplish more within any usage limits.

Crafting Your First Real Prompt

Enough theory—let's get practical. Writing a good prompt is like giving directions: the clearer you are, the better the results.

The Anatomy of a Great Prompt

1. **Context:** Set the stage. "I'm planning a dinner party for 8 people this weekend."
2. **Specific Request:** What you need "I need help creating a menu that's impressive but not too complicated."

3. **Constraints or Preferences:** Your parameters "Two guests are vegetarian, and I have about 3 hours to cook."
4. **Desired Output:** How you want the answer "Please provide a menu with appetizer, main course, and dessert options, plus a cooking timeline."

Put it all together: "I'm planning a dinner party for 8 people this weekend. I need help creating a menu that's impressive but not too complicated. Two guests are vegetarian, and I have about 3 hours to cook. Please provide a menu with appetizer, main course, and dessert options, plus a cooking timeline."

See the difference between that and "help me plan dinner"?

Common Rookie Mistakes (And How to Avoid Them)

We all make mistakes. Here are the most common ones, so you can skip right past them:

Mistake #1: The Vague Request

Wrong: "Write something good." **Right:** "Write a 200-word professional bio for my LinkedIn profile. I'm a marketing manager with 5 years of experience in tech startups."

Mistake #2: The Information Dump

Wrong: [Pastes 10,000 words without any guidance]. **Right:** "Here's my report. Please summarize the key findings in 3 bullet points and identify any potential risks mentioned."

Mistake #3: The Mind Reader Expectation

Wrong: "You know what I mean." **Right:** Actually explaining what you mean (Claude's smart, not psychic)

Mistake #4: The One-Shot Wonder

Wrong: Expecting perfection on the first try. **Right:** Iterating and refining—"That's great! Can you make the tone more formal and add statistics?"

Building Momentum: Your Second and Third Prompts

Congratulations! You've sent your first prompt, and Claude has responded. Now what? This is where the real magic happens—in the follow-up.

The Art of the Follow-Up

Think of your conversation with Claude like sculpting: your first prompt creates the rough shape, and follow-ups refine it into something beautiful.

Good Follow-Ups:

- "That's helpful! Can you expand on point #3?"
- "Perfect tone, but can we make it 50 words shorter?"
- "I love this approach. Now let's apply it to my specific situation..."
- "Can you explain that technical term you used?"

Building on Previous Responses

Since Claude remembers your entire conversation within its context window, you can build complex projects step by step:

1. "Help me outline a blog post about sustainable gardening"
2. "Great outline! Now let's write the introduction"
3. "Perfect. Can you develop the first main point about composting?"
4. "Excellent. Let's add some statistics to support this section"

See how each prompt builds on the last? You're not starting over—you're constructing something together.

Making Claude Your Permanent Thinking Partner

Here's the mindset shift that separates casual Claude users from power users: stop thinking of Claude as a tool and start thinking of it as a thinking partner.

The Collaboration Mindset

Instead of "Claude, do this for me"; Think: "Claude, let's work through this together."

This subtle shift changes everything:

- You get better results
- You learn more in the process
- You maintain ownership of your work
- You develop better AI collaboration skills

Starting Your Next Conversation

When you're ready to start a new thread (maybe you want to switch topics completely or just start fresh), remember:

- Each new conversation starts with a blank slate
- Claude won't remember previous threads
- You can always copy key information from old conversations
- Starting fresh can sometimes lead to new insights

Real-World First Conversations

Let's look at some first conversations that led to great outcomes:

The Entrepreneur

First Prompt: "I have an idea for a meal-planning app. Can you help me think through the concept?" **Result:** Over the next hour, they developed a complete business model, identified target users, and created a feature list.

The Student

First Prompt: "I'm struggling to understand the difference between RAM and storage. Can you explain it simply?" **Result:** Not only understood the concept but went on to ace their computer science exam.

The Writer

First Prompt: "I want to write a mystery novel but don't know where to start. Help?"
Result: Developed a complete plot outline, character profiles, and wrote the first chapter.

Your First Conversation Checklist

Before you dive in, here's your pre-flight checklist:

- ☐ Clear idea of what you want to accomplish
- ☐ Specific first prompt ready
- ☐ Understanding that iteration is normal and good
- ☐ Realistic expectations (Claude's smart, not magic)
- ☐ Open mind about the possibilities
- ☐ Cup of coffee (optional but recommended)

The Bottom Line

Your first conversation with Claude doesn't have to be perfect. In fact, it probably won't be—and that's absolutely fine. What matters is that you start, experiment, and discover how this remarkable AI can enhance your work and life.

Remember: Claude is infinitely patient, endlessly helpful, and genuinely designed to make your life easier. Whether you're writing your first prompt right now or planning to start tomorrow, you're about to begin a collaboration that can transform how you think, work, and create.

So go ahead—say hello to Claude. Your AI thinking partner is waiting, and your first conversation is just the beginning of an incredible journey.

In the next chapter, we'll dive deeper into understanding Claude's capabilities and limitations, ensuring you get the most out of every interaction. But for now, focus on starting that first conversation. The future of human-AI collaboration begins with a single prompt.

Ready? Type away. Claude's listening.

CHAPTER 3

Understanding Claude's Capabilities and Limitations

In This Chapter

- Discovering Claude's superpowers (and why they matter to you)
- Understanding the boundaries: what Claude won't and can't do
- Learning to spot and handle AI hallucinations like a pro
- Navigating content policies without hitting roadblocks
- Making peace with knowledge cutoffs and working around them

You wouldn't use a hammer to paint a portrait or a paintbrush to drive nails. Understanding what Claude can and can't do is just as important as knowing how to use it. This chapter provides a comprehensive framework for evaluating Claude's capabilities and limitations, enabling you to set realistic expectations and work more efficiently. Think of it as your owner's manual for optimal AI collaboration—principles that will serve you whether you're using today's Claude or tomorrow's AI assistants.

R. Dickey, *Mastering Claude AI*, https://doi.org/10.1007/979-8-8688-2001-4_3

Claude's Superpowers: What Makes It Shine

Let's start with the good stuff. Claude has some genuinely impressive capabilities that can transform how you work, learn, and create. These abilities represent the cutting edge of AI development, though as with all technology, specific features continue to evolve.

Writing and Language Mastery

Claude doesn't just write—it understands the nuances of language in ways that would make your high school English teacher weep with joy. Here's what advanced language models can do:

> **Style Chameleon:** Need a formal business report? Done. Want a friendly email to your team? Easy. Looking for creative fiction with a noir twist? Claude's got you covered. Modern AI can adapt writing style with remarkable flexibility.
>
> **Grammar Guru:** Claude catches errors you'd miss after five proofreads. It understands not just rules but context—knowing when to break conventions for effect.
>
> **Multilingual Marvel:** While Claude works best in English, current language models can handle numerous languages with varying degrees of fluency. It's like having a polyglot friend who never shows off about it.

Analysis and Reasoning

This is where Claude really flexes its digital muscles:

> **Pattern Recognition:** Claude can spot trends and connections in data that would take humans hours to find. It's like having a detective who never needs coffee breaks.
>
> **Complex Problem Solving:** From debugging code to planning projects, Claude can break down complex problems into manageable steps. Think of it as your personal strategic consultant who never sends invoices.

Critical Thinking: Modern AI systems can evaluate arguments and identify logical fallacies, though they work best when errors are clearly problematic rather than systematically challenging all assumptions. Claude is refreshingly honest about uncertainty too.

Creative Collaboration

Creativity isn't just for humans anymore:

Idea Generation: Stuck on a project? Claude can generate dozens of approaches, angles, and solutions. It's like brainstorming with someone who's read everything and forgets nothing.

Storytelling: From plot outlines to character development, Claude can help craft narratives that actually make sense (looking at you, final season of Game of Thrones).

Problem Reframing: Sometimes the best solution is asking a different question. Claude excels at looking at problems from new angles.

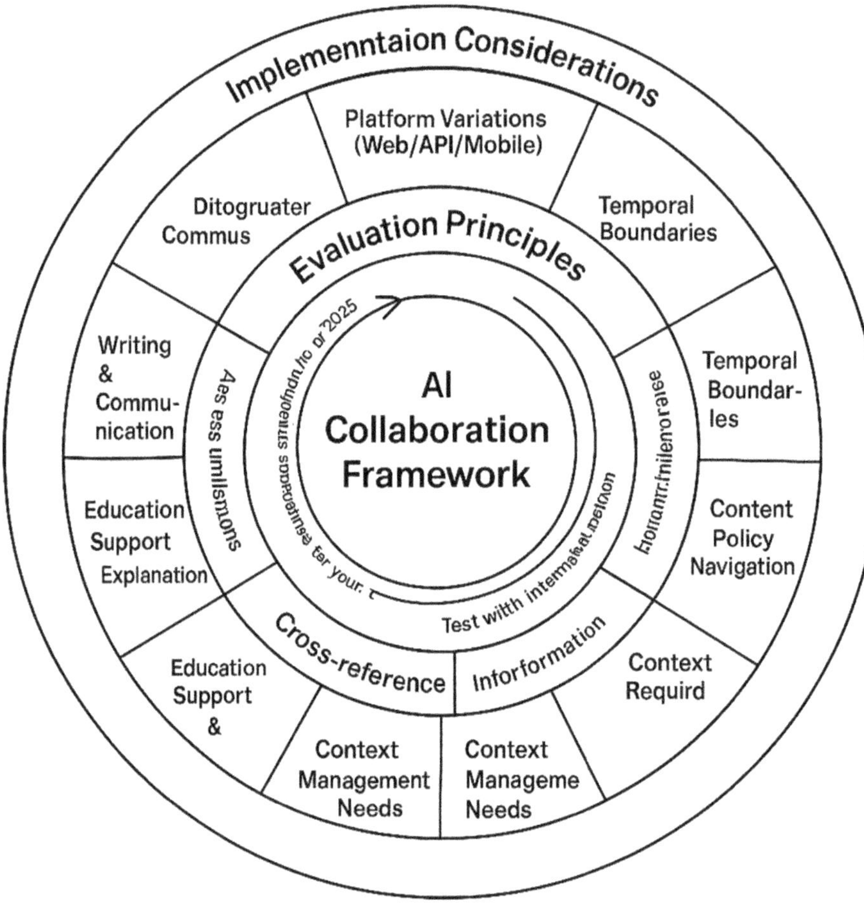

Figure 3-1. *AI Capability Evaluation Framework (2025): A comprehensive approach to understanding current AI system capabilities, limitations, and evaluation principles that remains valuable as technology evolves across platforms and implementations*

Educational Support

Claude might be the best study buddy you never had:

> **Adaptive Explanations:** Claude can explain quantum physics to a five-year-old or discuss advanced calculus with a mathematician. It adjusts explanations based on the language patterns you use, though you'll get better simplified explanations by using correspondingly simple language yourself.

Interactive Learning: Rather than just giving answers, Claude can guide you through problems, helping you understand the why, not just the what.

Knowledge Synthesis: Claude can pull together information from multiple domains, creating connections that deepen understanding.

Multimodal Capabilities: Beyond Text

As AI systems evolve, they're gaining capabilities beyond text processing. Current Claude models (as of 2025) include:

Visual Analysis: Claude can analyze photographs, charts, diagrams, and other images to extract information and provide insights. This represents a significant evolution in AI capability.

Document Processing: Modern versions can handle various file types including PDFs, Word documents, and spreadsheets. The specific limits and supported formats continue to evolve, so verify current capabilities for your use case.

Chart and Graph Interpretation: Whether it's financial reports or research data, Claude can interpret visual elements and explain trends or patterns—a capability that exemplifies how AI is moving beyond pure text processing.

Current Information Access

One traditional AI limitation—knowledge cutoffs—is being addressed through evolving capabilities:

Web Integration: Recent developments allow Claude to search for current information, helping overcome training data limitations. This represents an important shift in how AI systems handle real-time information.

Dynamic Research: Rather than being limited to training data, Claude can now find recent developments, current prices, and breaking news when configured with appropriate tools.

Note Information access capabilities vary by platform and implementation. Always verify what features are available in your specific use case.

Code and Data Processing

Technical capabilities continue expanding:

Code Execution: In appropriate environments, Claude can run code for data processing and analysis—moving beyond just writing code to actually executing it.

File Analysis: Beyond reading documents, Claude can process spreadsheets, analyze data files, and work with technical formats.

Tool Integration: Through APIs and integrations, Claude can interact with external systems, though this depends heavily on specific implementations.

The Reality Check: Understanding Limitations

Now for the part where we establish realistic expectations. Claude is impressive, but it's not omniscient, omnipotent, or able to make your coffee (yet). Understanding these limitations helps you work more effectively and avoid frustration—principles that apply to any AI system.

Knowledge Boundaries: Training Data Realities

All AI systems work within the boundaries of their training data. For Claude, this means:

Training Data Cutoffs: Claude's foundational knowledge reflects information available during training. While current systems include data through early 2025, the specific cutoff varies by model version.

What This Means:

- Core knowledge doesn't include very recent events or developments
- Training data doesn't cover brand-new technologies or trends
- Historical and established knowledge remains extensive and valuable

Working Around It:

- Use available search capabilities for current information
- Be specific about timeframes when asking questions
- Provide context about recent events if they're relevant
- Leverage Claude's strong foundational knowledge for established topics

The Hallucination Problem: When AI Generates Fiction

Let's talk about one of AI's persistent challenges: hallucinations. In AI terms, a hallucination is when the model generates information that seems plausible but is actually false. This is a fundamental limitation of current language model technology.

Common Triggers:

- Asking about very specific, obscure information
- Requesting details about fictional scenarios as if they were real
- Pushing for information the system doesn't have
- Complex calculations without showing work

Detection Strategies:

- If something seems too specific or convenient
- When exact figures appear without sources
- If information seems inconsistent with known facts
- When details change between responses

Your Defense Strategy:

- Always fact-check important information
- Ask for reasoning and sources
- Cross-reference critical data
- Understand that confident-sounding responses may still be incorrect

This is a fundamental characteristic of current AI technology—impressive capability coupled with the potential for confident-sounding errors.

Content Policies: Ethical Boundaries

Responsible AI systems have built-in boundaries, and that's actually a good thing. Content policies ensure AI remains helpful without being harmful.

Common Restrictions:

- Illegal activities (yes, even hypothetically)
- Harmful content creation
- Deceptive practices
- Inappropriate content
- Professional advice requiring licenses

Why These Limits Exist: These boundaries reflect broader societal decisions about responsible AI development. They're designed to ensure AI technology helps humanity without enabling harm.

Working Within Boundaries:

- Reframe requests constructively
- Focus on educational or creative purposes
- Understand the principles behind policies
- Remember that legitimate uses are fully supported

Platform and Implementation Variables

An important reality: AI capabilities vary significantly based on how you access them:

Access Method Variations:

- Web interface vs. mobile vs. API implementations
- Free vs. paid subscription levels
- Geographic availability and regional rollouts
- Third-party integrations vs. direct access

Model Version Differences:

- Different Claude variants (Haiku, Sonnet, Opus) have varying capabilities
- Continuous updates and improvements
- Feature rollouts happening at different paces

The Practical Impact: Always verify current capabilities for your specific access method and subscription level. What works for one user may not be available to another.

Technical and Practical Boundaries

Understanding current limitations helps set appropriate expectations:

Processing Constraints: Even powerful AI systems have computational limits that affect response time and complexity handling.

Context Management: While context windows have grown substantially, extremely long conversations or massive documents may require active management.

File and Format Limitations: Current systems support many file types but have size limits and format restrictions that vary by platform.

Session Memory: AI systems don't maintain memory between separate conversations, though they excel at maintaining context within a single session.

The Art of Verification: Trust but Verify

Here's a principle for the AI age: AI systems are powerful tools that require verification for critical applications. This isn't about distrust—it's about responsible use.

When to Verify

Always Double-Check:

- Medical or health information
- Legal or financial advice
- Historical dates and figures
- Scientific data or statistics
- Any information for publication or critical decisions

Generally Reliable:

- General explanations and concepts
- Creative writing and brainstorming
- Programming syntax and logic
- Language and grammar assistance
- Mathematical principles

Verification Strategies

The Smart Approach:

1. Ask for reasoning and sources
2. Cross-reference important facts
3. Use multiple sources for critical information
4. Pay attention to confidence levels in responses
5. When in doubt, consult authoritative sources

Red Flags to Watch For:

- Oddly specific information without context
- Claims that seem too convenient
- Information contradicting established knowledge
- Evasive or overly vague responses

Working with Current Information

While AI training data has cutoffs, modern systems increasingly include ways to access current information:

When Current Information Helps

Ideal Applications:

- Recent news and current events
- Updated statistics and data
- New technological developments
- Real-time information needs
- Fact-checking and verification

Continuing Limitations:

- Quality depends on search results
- Paywalled content may be inaccessible
- Processing time varies
- Results still require verification

Best Practices

- Be explicit about needing current information
- Provide context about what you're seeking

- Remember that even current results need verification
- Use real-time access to complement, not replace, foundational knowledge

Real-World Application Scenarios

Let's examine how these capabilities and limitations play out across different use cases:

Scenario 1: Business Intelligence

AI Strengths:

- Structuring and organizing reports
- Analyzing provided data and documents
- Creating clear, professional content
- Generating insights from uploaded files
- Researching current market conditions when equipped with search

Approach Carefully:

- Accessing proprietary company data requires explicit sharing
- Strategic business decisions need human judgment and context
- Real-time financial data depends on available information sources
- Company-specific culture and internal politics remain human domains

Scenario 2: Academic and Research Work

AI Strengths:

- Explaining complex concepts clearly
- Structuring arguments and papers
- Analyzing uploaded research documents

- Providing study strategies and learning support
- Finding current information for research projects

Important Limitations:

- Cannot replace original research or critical thinking
- Academic integrity policies vary by institution
- Access to specialized databases depends on configuration
- Citation and attribution remain your responsibility

Scenario 3: Creative Projects

AI Strengths:

- Generating ideas and creative concepts
- Developing characters and narrative structures
- Providing feedback and suggestions
- Overcoming creative blocks
- Analyzing visual references and inspiration

Creative Boundaries:

- Cannot replace human artistic vision and personal style
- Commercial success depends on many factors beyond AI assistance
- Visual creation capabilities vary significantly by platform
- Understanding audience and market requires human insight

Maximizing AI Collaboration

Success with AI comes from understanding both capabilities and limitations, then working strategically within those boundaries:

Strategic Approaches

Leverage Strengths:

- Use AI for brainstorming and ideation
- Delegate routine analysis and writing tasks
- Employ AI for learning complex topics
- Take advantage of document processing for research
- Utilize current information access when available

Compensate for Limitations:

- Verify important information through multiple sources
- Understand your platform's specific capabilities
- Maintain realistic expectations about AI assistance
- Keep human judgment central to important decisions

The Partnership Mindset

The most effective approach treats AI as a powerful collaborator rather than a replacement:

- Combine AI processing power with your creativity
- Merge AI knowledge with your experience
- Balance AI consistency with your intuition
- Align AI capabilities with your specific goals

Future-Proofing Your AI Skills

As AI capabilities continue evolving, focus on principles that will remain relevant:

Timeless Skills:

- Clear communication and prompt design
- Critical evaluation of AI outputs

- Effective verification and fact-checking
- Strategic thinking about when and how to use AI
- Understanding the relationship between AI capabilities and limitations

Adaptive Mindset:

- Stay informed about capability improvements
- Regularly reassess what AI can and cannot do
- Adjust workflows as features evolve
- Maintain healthy skepticism while embracing useful capabilities

The Bottom Line

Claude represents the current state of the art in AI assistance—powerful, versatile, and genuinely useful for a wide range of tasks. Like any sophisticated tool, it requires understanding both its capabilities and its limitations to use effectively.

The key insight isn't about memorizing what Claude can do today, but rather developing frameworks for evaluating and working with AI systems as they continue to evolve. The principles of clear communication, critical verification, and strategic collaboration will serve you well whether you're working with today's Claude or tomorrow's AI assistants.

Understanding these capabilities and limitations isn't about focusing on what AI can't do—it's about maximizing what it can do while maintaining appropriate oversight and verification. When you understand both the power and the boundaries, you can work more effectively and make better decisions about when AI assistance will be most valuable.

In the next chapter, we'll dive into the art and science of prompt engineering—how to communicate with Claude in ways that get you exactly what you need, every time. Because knowing what Claude can do is only half the battle; the other half is knowing how to ask for it effectively.

Ready to become a prompt engineering expert? Turn the page, and let's transform you from a Claude user into someone who can collaborate effectively with any AI system.

CHAPTER 4

The Art of Prompting: Getting Better Responses

In This Chapter

- Mastering the dark art of prompt engineering (it's easier than it sounds)
- Writing prompts that get you exactly what you want
- Learning the power of context, specificity, and examples
- Discovering advanced techniques like role-playing and few-shot learning
- Perfecting your prompts through iteration (aka the art of the do-over)

If Claude is a genie in a bottle, then prompts are your wishes—and we all know what happens with poorly worded wishes. This chapter transforms you from someone who types questions into Claude to a prompt engineering wizard who gets precisely what they need, every single time. No magic lamp required.

Prompt Engineering: The Most Important Skill You've Never Heard Of

Let's start with a truth bomb: The difference between frustration and amazement with Claude often comes down to how you ask for what you want. Prompt engineering is the art and science of crafting inputs that get you optimal outputs. Think of it as learning Claude's language.

R. Dickey, *Mastering Claude AI*, https://doi.org/10.1007/979-8-8688-2001-4_4

What Is Prompt Engineering, Really?

Prompt engineering sounds like something that requires a hard hat and a degree from MIT. In reality, it's simply the skill of communicating effectively with AI. It's the difference between

Vague: "Write something about dogs."

Engineered: "Write a 300-word blog post about the benefits of adopting senior dogs, targeting potential first-time dog owners. Use a warm, encouraging tone and include three specific health benefits."

See the difference? One is throwing spaghetti at the wall. The other is serving a perfectly plated meal.

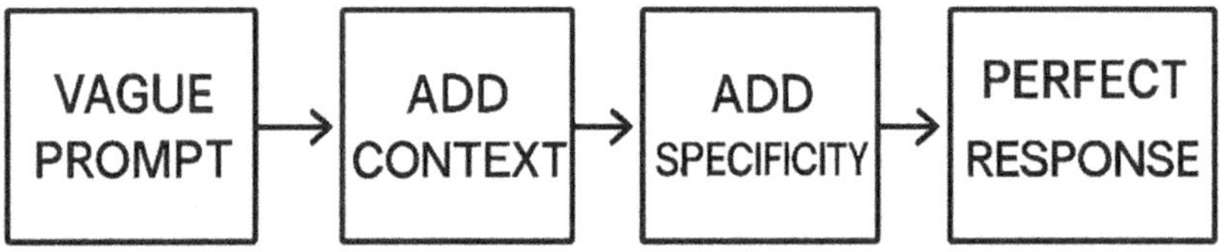

Figure 4-1. *The prompt engineering flow: Each step adds clarity and specificity, transforming vague requests into perfect responses. Like building blocks, each addition strengthens your prompt's effectiveness*

Why Your Prompts Matter More Than You Think

Here's what good prompt engineering gets you:

Precision: Exactly what you need, not sort-of-maybe-close

Efficiency: Less back-and-forth, more getting things done

Consistency: More reliable results you can build upon

Creativity: Better ideas through better questions

Learning: Understanding how AI responds by observing patterns in successful prompts

Bad prompts are like playing telephone with someone wearing earmuffs. Good prompts are like having a clear conversation with someone who's genuinely trying to help.

Important Note Prompting techniques work most of the time, but not always. As you develop your skills, you'll learn to recognize when an approach isn't working and try alternative strategies.

The Anatomy of a Perfect Prompt

Let's dissect what makes a prompt work. Think of it as a recipe—leave out key ingredients, and your soufflé becomes a pancake.

The Essential Ingredients

1. **Context: The Background** Context gives Claude the big picture. It's the difference between walking into a movie halfway through and starting from the beginning.

   Without Context: "Fix this email." **With Context:** "I'm reaching out to a potential client after our initial meeting. The email below needs to be more professional while maintaining a friendly tone."

   Example of Context Impact:

   - **Vague Prompt:** "Write about productivity" ➤ Generic, unfocused response
   - **Contextual Prompt:** "Write about productivity for remote workers struggling with home distractions" ➤ Targeted, specific advice

2. **Specificity: The Details** Specificity is your best friend. Vague requests get vague responses. It's like ordering "something good" at a restaurant vs. ordering "grilled salmon with lemon butter sauce"—you'll get food either way, but only one gets you what you actually wanted.

   Vague: "Help me with my presentation." **Specific:** "Create a 5-slide outline for a 10-minute presentation on renewable energy for high school students. Include key statistics and one interactive element."

3. **Format: The Structure** Tell Claude how you want your response structured. It's remarkably good at following formatting instructions.

   Examples:

   - "Provide your response as a numbered list"
   - "Write this as a formal business letter"
   - "Format as a table with three columns"
   - "Create a step-by-step tutorial"

4. **Constraints: The Boundaries** Constraints aren't limitations—they're clarity. They help Claude understand exactly what you need.

   Types of Constraints:

   - **Length**: "Keep it under 200 words"
   - **Tone**: "Use a conversational, friendly tone"
   - **Audience**: "Explain for someone with no technical background"
   - **Scope**: "Focus only on solutions available in the US"

The Power of Examples: Few-Shot Learning

Here's where things get interesting. Few-shot learning is a technique where you show Claude examples of what you want, demonstrating the pattern rather than just describing it. It's like teaching by demonstration rather than explanation.

Zero-Shot (No Examples): "Write product descriptions for leather office chairs"

One-Shot (Single Example): "Write a product description like this example:

Modern Desk Lamp: 'Sleek aluminum meets warm LED light in this architectural statement piece. Adjustable arm reaches exactly where you need illumination, while the weighted base ensures stability during late-night inspiration sessions.'

Now write one for a leather office chair."

Few-Shot (Multiple Examples): "Write product descriptions in this style:

- **Oak Coffee Table**: 'Crafted from solid oak, this timeless piece brings warmth to any living room. The natural grain tells a story in every swirl.'
- **Pine Bookshelf**: 'Six generous shelves of sustainable pine offer a home for your literary adventures. Simple assembly, lasting beauty.'

Now write one for a maple dining table."

Claude will match the style, tone, and structure of your examples. It's like having a skilled collaborator who understands patterns rather than just copying them.

Advanced Prompting Techniques

Ready to level up? These techniques separate the prompt engineering rookies from the pros. Remember, these approaches often improve results, but effectiveness varies by task and context.

Role-Playing: Putting Claude in Character

Role-playing isn't just for Dungeons & Dragons. Asking Claude to adopt a specific role or perspective can dramatically improve responses by providing context for tone, expertise level, and approach simultaneously.

Basic Prompt: "Explain quantum computing"

Role-Playing Prompt: "You're a professor of physics known for using creative analogies. Explain quantum computing to journalism students using metaphors from everyday life."

Other Effective Roles:

- "You're a patient elementary school teacher..."
- "Act as an experienced business consultant..."
- "You're a friendly but detail-oriented editor..."
- "Respond as a creative writing coach..."

Chain of Thought: Show Your Work

Chain of thought prompting asks Claude to explain its reasoning step by step. This technique is most effective when you need results with intermediate, verifiable steps rather than direct answers—particularly useful for complex problems that benefit from systematic breakdown.

Without Chain of Thought: "Is this business idea viable?"

With Chain of Thought: "Analyze this business idea by thinking through

1. Market demand
2. Competition analysis
3. Required resources
4. Potential challenges
5. Revenue potential

Then provide your overall assessment."

Important Caveat While chain of thought can improve results by forcing task decomposition into logical steps, Claude emulates reasoning patterns from its training data rather than performing true logical analysis. The reasoning may sound convincing but still requires verification, especially in domains where flawed reasoning patterns exist in training data.

Iteration: The Art of Refinement

Here's a secret: Professional prompt engineers rarely nail it on the first try. Iteration is the process of refining your prompts based on Claude's responses, and it's completely normal and valuable.

The Iteration Cycle:

1. **First Attempt:** Get something close to what you want
2. **Analyze:** What's working? What's missing?

3. **Refine:** Adjust your prompt based on the response
4. **Repeat:** Continue until you get closer to your goal

Example Iteration:

- **Version 1**: "Write about productivity"
- **Version 2**: "Write 5 productivity tips for remote workers"
- **Version 3**: "Write 5 unconventional productivity tips for remote workers, focusing on mental health. Include a brief example for each."

Each iteration gets you and Claude closer to what you need.

The Meta-Prompt: Asking Claude to Help with Prompts

Here's a pro move: Ask Claude to help you write better prompts. It's meta, but it works.

"I want to get help with [your goal]. What information would you need from me to provide the best possible assistance? What details should I include in my prompt?"

Claude can help you identify what context, constraints, and specifics would be most helpful.

Common Prompting Pitfalls (and How to Avoid Them)

Even experienced users fall into these traps. Learn from their mistakes and develop strategies for when techniques don't work as expected:

Pitfall #1: The Kitchen Sink Approach

> **Wrong:** Dumping everything into one massive prompt. **Right:** Breaking complex requests into manageable steps

Instead of asking Claude to write, edit, format, and optimize a report all at once, tackle each step separately.

Pitfall #2: The Mind Reader Fallacy

> **Wrong:** "You know what I mean." **Right:** Explicitly stating your requirements

Claude is smart, but it can't read between lines that aren't there.

Pitfall #3: The One-Size-Fits-All Prompt

> **Wrong:** Using identical prompt structures for every request. **Right:** Adapting your approach to the task

While most effective prompts follow a basic template of context + goal + format, different tasks require different emphases and implementation approaches. Writing code requires different prompting considerations than writing poetry.

Pitfall #4: The Perfectionist Paralysis

> **Wrong:** Trying to craft the perfect prompt before starting. **Right:** Starting simple and iterating

Your first prompt is a draft, not a final submission.

Pitfall #5: Over-Constraining

> **Wrong:** Providing so many specific criteria that they conflict. **Right:** Relevant specificity that works together
>
> **Remember**: relevant specificity beats vagueness, but multiple specific criteria can create unrealistic or conflicting constraints.

Real-World Prompting Scenarios

Let's see these principles in action across different use cases:

Scenario 1: The Email Makeover

Poor Prompt: "Make this email better"

Better Prompt: "Rewrite this email to my manager requesting time off. Make it more professional while keeping it brief. Maintain a respectful but confident tone. The key points to preserve are dates (July 15-22), reason (sister's wedding), and coverage plan (Jake will handle my projects)."

Scenario 2: The Code Debug

Poor Prompt: "Fix my code"

Better Prompt: "This Python function should calculate compound interest but returns incorrect values. Please

1. Identify the error
2. Explain why it's happening
3. Provide the corrected code
4. Add comments explaining the fix"

Scenario 3: The Creative Brief

Poor Prompt: "Give me marketing ideas"

Better Prompt: "Act as a creative director. Generate 5 innovative marketing campaign concepts for a sustainable water bottle brand targeting college students. For each concept, include

- Campaign name
- Main message
- Primary platform
- One unique activation idea"

Building Your Prompting Toolkit

Success with Claude comes from developing a toolkit of prompting techniques and learning to recognize patterns in what works:

Your Prompt Templates

Start collecting templates for common tasks and note which approaches work best for different types of requests:

Research Template: "Research [topic] and provide:

- Overview (2-3 sentences)
- Key findings (5 bullet points)
- Controversies or debates
- Recent developments
- Reliable sources for further reading"

Writing Template: "Write a [type of content] about [topic] for [audience].

- Length: [word count]
- Tone: [formal/casual/etc.]
- Key points to include: [list]
- Avoid: [things to skip]"

Problem-Solving Template: "Help me solve [problem]. Please:

1. Clarify the problem
2. Identify potential causes
3. Suggest solutions with pros/cons
4. Recommend the best approach
5. Outline implementation steps"

The Prompt Engineering Checklist

Before hitting send, ask yourself:

- ☐ Have I provided enough context?
- ☐ Is my request specific and clear?
- ☐ Did I specify the format I want?
- ☐ Are my constraints reasonable and compatible?
- ☐ Would examples help clarify my need?
- ☐ Am I asking for one thing at a time?
- ☐ Do I have a backup approach if this doesn't work?

The Art of Conversational Prompting

Remember, you're having a conversation, not programming a computer. The best prompting often feels natural:

Building on Responses

> **First Prompt:** "What are the main challenges in sustainable agriculture?"
>
> **Follow-Up:** "Interesting. Can you elaborate on the water scarcity issue you mentioned? Specifically, what solutions are showing promise?"
>
> **Further Refinement:** "Let's focus on drip irrigation. Can you explain how it works and why it's effective, as if you're talking to farmers who are skeptical about new technology?"

Each prompt builds naturally on the last, creating a collaborative dialogue.

The Power of "Why" (with Caveats)

Don't be afraid to ask Claude to explain its reasoning, but remember that like humans, Claude can provide flawed reasoning and may reason backwards from conclusions. This approach is most effective in domains where good reasoning patterns exist in training data:

- "Why did you recommend that approach?"
- "What factors did you consider?"
- "What alternatives did you rule out?"

Understanding Claude's logic helps you prompt better in the future, but always verify the reasoning independently.

Developing Your Prompting Skills Over Time

Becoming proficient at prompting is a learning journey that happens across multiple conversations:

The Learning Process

Weeks 1-2: Basic Structure

- Focus on clear context and specific requests
- Practice the basic prompt anatomy
- Learn to iterate rather than expecting perfection

Month 1: Pattern Recognition

- Notice which approaches work best for different tasks
- Start building your personal prompt library
- Develop fallback strategies for when techniques fail

Ongoing: Advanced Techniques

- Experiment with role-playing and chain of thought
- Recognize when to simplify overly complex approaches
- Build intuition for matching techniques to tasks

Building Your Personal Prompt Library

Over time, collect and refine prompts that work well for your common tasks:

- Templates for recurring work activities
- Successful role-playing scenarios
- Effective few-shot examples for your domain
- Backup approaches for when your first attempt fails

From Novice to Master: Your Prompting Journey

Becoming a prompt engineering master doesn't happen overnight. It's a journey of experimentation and pattern recognition:

Novice: "Help me write." **Intermediate:** "Write a 500-word blog post about remote work productivity." **Advanced:** "Write a 500-word blog post about remote work productivity for busy parents. Use a conversational tone, include one personal anecdote, three actionable tips, and end with an encouraging call-to-action. Avoid generic advice like 'make a schedule.'" **Master:** *Achieves the above through natural conversation, adapting techniques based on responses, and knowing when to try different approaches*

The Bottom Line

Prompt engineering isn't about tricking Claude or finding secret commands. It's about clear communication, thoughtful structure, and iterative refinement. The best prompts help you and Claude work together more effectively, though results will vary and techniques sometimes fail.

Remember these key principles:

- Context enables better responses
- Relevant specificity beats vagueness (but avoid over-constraining)
- Examples demonstrate patterns more clearly than descriptions
- Iteration is normal, valuable, and expected
- Role-playing provides useful context and perspective

- Different tasks benefit from different approaches
- Developing prompting skills is an ongoing learning process

The beauty of prompt engineering is that you get better with practice. Every interaction teaches you something new about how to communicate with AI effectively, building your intuition for what works when.

As we close Part I of this book, you now have the foundation: You understand what Claude is, how to start conversations, what it can and can't do, and how to ask for what you want. You're ready to transition from a basic user to a power user.

In Part II, we'll dive into specific applications—from writing and editing to coding and analysis. But everything builds on the prompt engineering skills you've learned here. Master these fundamentals, and you'll be amazed at what you and Claude can accomplish together.

Ready to put these skills to work? Part II awaits, where we'll transform theory into practice across every area of your life and work. The conversation is just beginning.

PART II

Practical Applications

CHAPTER 5

Writing and Communication Mastery

In This Chapter

- Turning Claude into your personal writing assistant (goodbye, writer's block!)
- Mastering every type of writing from emails to novels
- Learning the secret sauce of collaborative editing
- Discovering how to maintain your voice while leveraging AI
- Avoiding the pitfalls that scream "AI wrote this!"

Remember that English teacher who bled red ink all over your essays? Well, Claude is like having that teacher's knowledge combined with your cool friend's encouragement and a professional editor's skills—minus the judgmental sighs. This chapter transforms you from someone who stares at blank pages to someone who makes blank pages surrender. Whether you're crafting the perfect email, writing the Great American Novel, or just trying to sound intelligent in your team's Slack channel, Claude's got your back.

Your AI Writing Partner: More Than Just Spell Check

Let's get one thing straight: Claude isn't here to replace your **voice** (that's your unique writing style and personality that makes your writing sound like you, not like a robot or someone else). Think of it more like having a ridiculously talented writing buddy who's

R. Dickey, *Mastering Claude AI*, https://doi.org/10.1007/979-8-8688-2001-4_5

available 24/7, never gets tired of your questions, and somehow knows everything about everything. It's the difference between writing alone in a cave and having a collaborative partner who makes everything better.

A Framework for AI-Assisted Writing

Before diving into specific techniques, let's establish a clear framework for how writing principles connect to Claude collaboration:

Step 1: Define Your Writing Goal

- What type of writing? (email, report, creative piece)
- Who's your audience?
- What's your desired outcome?

Step 2: Leverage Claude's Strengths

- Use Claude for brainstorming and structure
- Get help with tone and style adaptation
- Receive feedback and suggestions for improvement

Step 3: Maintain Your Voice

- Provide examples of your preferred style
- Give specific direction about tone and approach
- Always review and personalize Claude's suggestions

Step 4: Iterate and Refine

- Use Claude's feedback to improve drafts
- Ask for specific types of revision
- Build on responses through conversation

What Makes Claude Different from Other Writing Tools

You've probably used spell checkers, grammar tools, and maybe even those "make your writing better" apps that turn everything into corporate buzzword soup. Claude is fundamentally different:

Traditional Tools:

- Fix mistakes after you make them
- Apply rigid rules regardless of context
- Make everyone sound the same
- Focus on what's wrong

Claude:

- Helps you avoid mistakes in the first place
- Understands context and nuance
- Preserves and enhances your unique voice
- Focuses on making your writing better

It's like the difference between a GPS that only tells you when you've made a wrong turn vs. one that helps you plan the best route from the start.

Understanding Tone: The Secret Ingredient

Tone in writing is like seasoning in cooking—it's the attitude or emotion conveyed in your words that determines how they "taste" to readers. Is your email friendly or formal? Is your blog post serious or playful? Claude can help you nail the right tone every time. Think of tone as the emotional flavor behind your words. A single message can be delivered in dozens of different tones:

- **Formal**: "I would appreciate your response at your earliest convenience."
- **Friendly**: "Hey, when you get a chance, could you let me know?"
- **Urgent**: "Need your answer ASAP—the deadline is tomorrow!"
- **Apologetic**: "Sorry to bother you, but I was wondering if you could help..."

Claude acts as your tone consultant, helping you match your message to your audience and purpose.

The Art of Audience Adaptation

Audience adaptation—adjusting your writing style, tone, and content to suit your intended readers—is crucial for effective communication. It's the difference between speaking to a five-year-old about quantum physics (lots of analogies involving toys) and presenting to physicists (bring on the equations). Claude can help you master this art by:

- Identifying your target audience's knowledge level
- Suggesting appropriate vocabulary and complexity
- Adjusting formality based on context
- Highlighting jargon that might confuse readers
- Recommending examples that resonate

Whether you're writing for tech-savvy millennials or traditional executives, Claude helps you speak their language.

Email Excellence: From "Per My Last Email" to Persuasion Pro

Let's start with the writing we usually do most often: emails. Between what you send and receive, you're likely dealing with dozens of emails daily. That's a lot of opportunities to sound brilliant—or like someone who shouldn't be allowed near a keyboard.

The Anatomy of the Perfect Email

Claude can help you nail every element through strategic **prompting** (the art of asking Claude for exactly what you need in a way that gets you the best results):

> **Subject Lines That Get Opened:** Instead of "Meeting," Claude helps you write, "Quick 15-min sync on Q4 budget—decision needed by Friday"

Openings That Connect: Instead of "Hope this email finds you well," Claude suggests, "Thanks for your insights in yesterday's meeting—your point about customer retention really resonated."

Clear, Actionable Content: Claude helps you structure emails that people actually want to read:

- Lead with the main point
- Use bullet points for multiple items
- Include clear next steps
- End with a specific call to action

How to Use Claude for Email Improvement

Prompt Framework for Email Enhancement:

"I need help with this email to [audience]. My goal is [specific outcome]. The tone should be [formal/friendly/urgent]. Here's my draft: [paste email]. Please help me improve [specific aspect: clarity/persuasiveness/brevity]."

Email Scenarios and Solutions

Scenario 1: The Delicate Follow-Up You: "Claude, I need to follow up with a client who hasn't responded to my proposal for two weeks. I want to be persistent but not annoying."

Claude helps craft: "Hi Sarah, I know you're juggling multiple priorities right now. I wanted to check if you had any questions about the proposal I sent on October 15th. If the timing isn't right, I'm happy to circle back when it works better for you. Would a quick 10-minute call help clarify anything?"

Scenario 2: The Bad News Email You: "Help me tell my team that the project deadline is moving up by two weeks."

Claude transforms bad news into action: "Team, I just received word that our launch date needs to move to March 1st. I know this is aggressive, but I believe in our ability to deliver. Let's meet tomorrow at 10 AM to revise our timeline and identify any resources you need to succeed. I'll bring coffee and donuts—we're in this together."

Scenario 3: The "I Need a Favor" Email You: "I need to ask my overworked colleague to review my presentation."

Claude creates reciprocity: "Hey Marcus, I know you're swamped with the product launch. Would you have 20 minutes this week to review my sales presentation? Your insights on customer pain points would be invaluable. Happy to return the favor—I could take that vendor call off your plate on Thursday if that helps."

Long-Form Content: From Blog Posts to Books

Now let's tackle the beast that is **long-form content** (any written piece over 1,000 words, like blog posts, articles, reports, white papers, or even books—basically anything longer than your typical email or social media post). Whether you're writing blog posts, reports, or that novel you've been "meaning to start," Claude is like having a writing coach, research assistant, and editor rolled into one.

The Blog Post Blueprint

Great blog posts follow a structure, and Claude helps you nail it every time using the power of **iteration** (the process of refining and improving your writing through multiple rounds of revision—think of it as sculpting, where each pass makes your writing better).

How to Use This Template with Claude:

This blueprint serves as both prompt guidance and evaluation criteria. Use it to:

1. **Initial Development:** Ask Claude to help you create each section systematically
2. **Evaluation:** Check your draft against these elements
3. **Iterative Improvement:** Request specific improvements for each component

The Hook (First 100 Words):

- Start with a story, statistic, or question
- Make a bold claim or challenge assumptions
- Promise value clearly
- Create curiosity

The Journey (Middle Sections):

- Break complex ideas into digestible chunks
- Use subheadings as mini-hooks
- Include examples and analogies
- Maintain momentum with varied sentence structure

The Landing (Conclusion):

- Summarize key takeaways
- Provide actionable next steps
- End with a thought-provoking question
- Include a clear call to action

Case Study: From Blank Page to Published Post

Let's watch Claude help transform a vague idea into a compelling blog post using collaborative brainstorming:

> **Writer**: "I want to write about productivity, but that feels too generic."
>
> **Claude**: "Let's narrow it down. What specific productivity challenge do you or your readers face most?"
>
> **Writer**: "Probably staying focused with all the distractions from notifications."
>
> **Claude**: "Great angle. How about: 'The 3-2-1 Method: How I Reclaimed 2 Hours of Deep Work Daily (Without Throwing My Phone in a Lake)'—personal, specific, and promises a concrete benefit."
>
> **Writer**: "I like it! But I don't know where to start."

> **Claude**: "Start with a relatable scenario. How about 'It was 11:47 AM, and I'd already checked my email 23 times. Not because I was expecting anything important—just because that little notification badge is apparently more addictive than potato chips at midnight.'"

See how Claude helps find the specific angle, create an engaging title, and start with a hook that readers can relate to?

Business Writing That Actually Gets Read

Corporate communication doesn't have to sound like it was written by a committee of robots who learned English from tax forms. Claude helps you write business documents that are professional AND engaging.

The Executive Summary That Executives Actually Read

An **executive summary** provides a comprehensive overview of complex reports and can serve as a standalone document—it's your document's elevator pitch and often the only section busy executives will read. Think of it as the movie trailer for your report.

> **Traditional Approach**: "This document provides a comprehensive analysis of various strategic initiatives designed to optimize operational efficiency across multiple organizational verticals..."
>
> **Claude's Approach**: "We can cut costs by 23% and increase customer satisfaction—here's how. This report outlines three specific changes we can implement within 60 days."

Proposals That Win

Claude can help you structure proposals that get to "yes" using **role-playing** techniques (asking Claude to respond from a specific perspective or persona, like "act as a skeptical CFO" or "respond as a potential customer"):

Effective Proposal Structure:

The Problem (Why Should They Care?): Don't just state facts—tell a story about impact

The Solution (What You're Proposing): Lead with benefits, follow with features

The Proof (Why You?): Use specific examples and metrics, not generic claims

The Plan (How It Works): Clear timeline, defined milestones, obvious next steps

The Investment (What It Costs): Frame as ROI, not expense

Making Data Sing

Got a report full of numbers? Claude helps you turn spreadsheet nightmares into compelling narratives:

> **Example of Improvement:** Poor: "Q3 revenue was $2.3M, representing a 15% increase over Q2."
>
> **Better**: "We broke the $2 million barrier this quarter—a 15% jump that puts us three months ahead of our growth targets. Here's what's driving the momentum..."

Note This example demonstrates the difference between data reporting and data storytelling.

Creative Writing: Unleashing Your Inner Novelist

Whether you're writing the next bestseller or just trying to make your company newsletter less soul-crushing, Claude is your creative collaborator.

Character Development That Breathes

Creating believable characters is hard. Claude helps by analyzing effective character development techniques rather than using **few-shot learning** (showing Claude examples of what you want so it can follow the pattern) and asking the right questions:

- What does your character want more than anything?
- What lie do they believe about themselves?
- What would they never do—until they have to?
- What small habit reveals their true nature?

Dialogue That Doesn't Make Readers Cringe

Example for Reference:

> Poor dialogue: "Hello, John. How are you today? I am fine. The weather is nice."
>
> Effective dialogue: "John slouched against the doorframe. 'You're late.' Sarah didn't look up from her laptop. 'You're observant.'"

See the difference? Real people don't speak in complete sentences or say exactly what they mean.

Plot Development Without Holes

Claude Can Help You:

- Identify plot inconsistencies before readers do
- Develop compelling subplots that enhance the main story
- Create authentic obstacles (not just random bad luck)
- Build satisfying resolutions that feel earned

The Editing Revolution: From First Draft to Final Polish

Here's a truth bomb: First drafts are supposed to suck. That's why they're called first drafts, not "perfect-on-the-first-try drafts." Claude makes the revision process less painful and more productive through systematic **editing**—reviewing and improving written content for clarity, accuracy, and effectiveness.

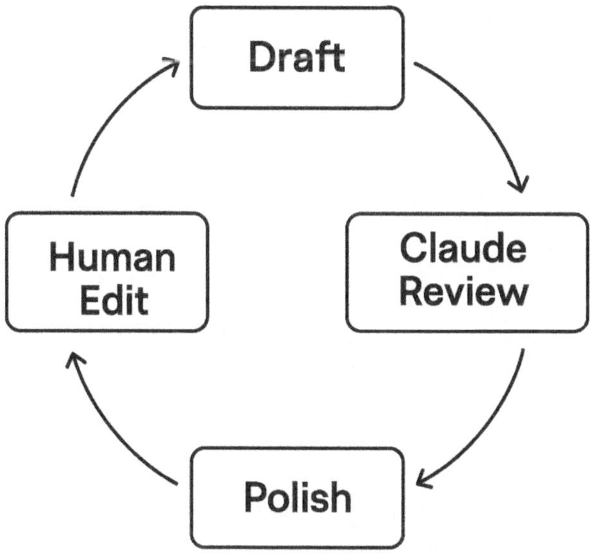

Figure 5-1. *The collaborative writing cycle: Human creativity and Claude's capabilities create a continuous feedback loop, with each stage improving the final output*

The Three-Pass Editing System

The **three-pass editing** system is a methodical approach to revision where you focus on different aspects in each round, like a sculptor working from rough shape to fine details.

How to Implement This System with Claude:

Pass 1: Structure and Flow Claude Prompt: "Please review this draft for overall structure and flow. Does the piece achieve its purpose? Is information in logical order? Are transitions smooth?"

Focus Areas:

- Does the piece achieve its purpose?
- Is information in the most logical order?
- Do transitions guide readers smoothly?

Pass 2: Clarity and Precision Claude Prompt: "Help me improve clarity and precision. Which complex ideas can be simplified? What unnecessary words should be cut? What jargon needs explanation?"

Focus Areas:

- Can complex ideas be simplified?
- Are there unnecessary words to cut?
- Is jargon explained or eliminated?

Pass 3: Polish and Proofreading Claude Prompt: "Please help with final polish and proofreading. Check for typos, grammar issues, consistency, and awkward phrasing."

Proofreading—carefully checking text for errors in spelling, grammar, and punctuation—is the final crucial step that separates professional writing from amateur work. This pass includes:

- Checking for typos and spelling errors
- Verifying grammar and punctuation
- Ensuring consistent formatting
- Confirming names and facts are correct
- Reading aloud to catch awkward phrasing

Common Writing Crimes and How to Fix Them

Crime #1: Passive Voice Overload Before: "The report was written by the team." After: "The team wrote the report." Claude helps you spot and fix passive constructions that drain energy from your writing.

Crime #2: Buzzword Bingo Before: "We're leveraging synergies to optimize stakeholder value." After: "We're working together to deliver better results." Claude translates corporate speak into human language.

Crime #3: The Wall of Text Before: [One massive paragraph that makes readers' eyes glaze over] After: Transform intimidating blocks into inviting paragraphs. When readers see shorter chunks, they think, "I can read this" instead of "I'll read this later" (which means never).

Maintaining Your Voice While Using AI

Here's the big fear: "Will using Claude make me sound like everyone else?" Short answer: Only if you let it. Long answer: Voice preservation requires active user effort and careful prompting, not automatic AI assistance. Claude enhances your **style**—the distinctive way you write and express yourself—but doesn't replace it.

The Voice Preservation Protocol

Voice preservation is the art of maintaining your unique writing style while using AI assistance—like keeping your accent when learning a new language. Here's how to implement this systematically:

Step 1: Know Your Voice Before using Claude, write a paragraph about something you're passionate about. This is your voice baseline.

Step 2: Direct the Collaboration Don't just accept Claude's first suggestion. Instead, use specific prompts:

- "Make this more conversational"
- "Add some humor here"
- "This needs more energy"
- "Keep my tendency to use pop culture references"

Step 3: The Final Polish Always do a final pass to ensure it sounds like you. Change words that you'd never use. Add your signature phrases. Make it yours.

Voice Examples in Action

Academic Voice: "The implications of this research suggest a paradigm shift in our understanding of user behavior."

Conversational Voice: "This research basically flips everything we thought we knew about users on its head."

Professional but Friendly: "Our research uncovered something surprising—users aren't behaving the way we expected, and that's actually good news."

Claude can help you maintain any of these voices consistently throughout your writing.

Advanced Techniques: The Power User's Toolkit

Ready to level up? These advanced techniques separate the Claude rookies from the pros.

The Systematic Iteration Method

Building on the basic iteration concepts discussed earlier, this approach provides more structured refinement:

1. **First Pass**: Get the basic structure and ideas
2. **Second Pass**: "Make this more compelling"
3. **Third Pass**: "Add specific examples"
4. **Fourth Pass**: "Punch up the opening and closing"

The Perspective Shift

Stuck? Ask Claude to approach from different angles:

- "How would a skeptic respond to this?"
- "What would a beginner need to know?"
- "What questions would an expert ask?"
- "How would I explain this to my grandmother?"

The Style Analysis Technique

Have a favorite writer? Claude can analyze their techniques for learning purposes, though it avoids direct imitation for copyright reasons:

- "What makes Hemingway's writing so punchy?"
- "How does Malcolm Gladwell structure his arguments?"
- "What techniques does Mary Roach use to make science funny?"

Then apply those insights to your own work.

Real Writers, Real Results

Let's look at how people might use Claude to level up their writing game. These are illustrative examples based on common user experiences:

> **Sarah, the Startup Founder**: "I used to spend hours agonizing over investor emails. Now Claude helps me nail the tone—professional but not stuffy. Last week, I got meetings with three VCs. My old emails got ignored."
>
> **Marcus, the Marketing Manager**: "Claude transformed my blog posts from SEO keyword soup into stories people actually want to read. Our engagement time doubled, and I'm writing three times faster."

Dr. Jennifer Chen: "I love research but hate writing papers. Claude helps me translate complex findings into clear narratives. My last paper got accepted on the first submission—that never happens!"

Tom the Technical Writer: "Documentation used to be torture. Claude helps me find analogies that make complex systems understandable. My docs went from most-avoided to most-requested."

The Writing Transformation Checklist

Ready to revolutionize your writing? Here's your action plan:

Week 1: Email Excellence

- ☐ Use Claude to rewrite your three most important email templates
- ☐ Practice the three-part structure (hook, body, action)
- ☐ Time how much faster you write emails

Week 2: Long-Form Focus

- ☐ Outline a blog post or article with Claude
- ☐ Write the first draft using the blueprint method
- ☐ Use the three-pass editing system

Week 3: Find Your Voice

- ☐ Write your voice baseline paragraph
- ☐ Experiment with different tones
- ☐ Create your personal style guide

Week 4: Advanced Applications

☐ Try the systematic iteration method on important documents

☐ Use perspective shifts for stuck projects

☐ Develop templates for recurring writing tasks

Common Pitfalls and How to Avoid Them

Even with Claude's help, writers stumble. Here's how to avoid the most common traps:

Pitfall #1: The Copy-Paste Trap

The **copy-paste trap** is when you blindly copy Claude's suggestions without customizing them—like wearing someone else's clothes without tailoring. Don't just copy Claude's suggestions verbatim. Always customize and personalize.

Pitfall #2: Losing Your Voice

If everything starts sounding the same, you're over-relying on Claude. Use it as a collaborator, not a ghostwriter.

Pitfall #3: Skipping the Human Touch

Claude is brilliant, but it doesn't know your audience like you do. Always apply your human judgment.

Pitfall #4: First Draft Syndrome

Just because Claude helps you write faster doesn't mean you should skip editing. Great writing is rewriting—even with AI help. The book you're reading is a testament to this concept.

The Bottom Line

Writing with Claude isn't about replacing your skills—it's about amplifying them. Think of it as upgrading from a bicycle to an e-bike. You're still pedaling, still choosing the direction, still enjoying the ride. You just get there faster and arrive less exhausted.

Whether you're crafting emails that get responses, blogs that get shared, or novels that get published, Claude is your writing force multiplier. It handles the heavy lifting so you can focus on what matters: your ideas, your voice, and your message.

Remember: The goal isn't to write like an AI. The goal is to write like the best version of yourself, with an AI assistant that never judges your split infinitives or rolls its eyes at your dad jokes.

In the next chapter, we'll explore how Claude can transform you into a data analysis wizard—even if spreadsheets currently make you break out in hives. But for now, go forth and write. Your blank pages don't stand a chance.

Ready to make every word count? Let's write.

CHAPTER 6

Research and Analysis Like a Pro

In This Chapter

- Transforming from Google searcher to research ninja
- Mastering the art of source verification (fake news, beware!)
- Learning to synthesize information like a think tank analyst
- Creating summaries that actually capture what matters
- Building research workflows that save hours (or days)

Research used to be about finding information. Now it's about making sense of too much information. That's where Claude transforms you from someone overwhelmed by options into someone who understands exactly what matters. Whether you're investigating competitors, exploring new topics, fact-checking claims, or diving deep into your latest obsession, Claude turns information chaos into clear insights. Think of it as the difference between having a library card and having a brilliant librarian who's read every book.

Important Note While Claude can be an excellent research collaborator, all AI-generated analysis should be verified against original sources. AI may identify patterns that don't exist or miss important contradictions in source material. This chapter provides frameworks for effective human-AI research collaboration while maintaining appropriate oversight.

R. Dickey, *Mastering Claude AI*, https://doi.org/10.1007/979-8-8688-2001-4_6

The Research Revolution: A Framework for AI-Assisted Investigation

Let's be honest: effective information management is the real challenge in modern research. The internet gives us unprecedented access to information, but knowing how to navigate, evaluate, and synthesize that information systematically is what separates casual searching from professional research.

Claude transforms research from a scavenger hunt into a guided expedition. Instead of drowning in information, you'll learn to navigate it systematically. But first, let's understand what real **research methodology** means in the AI age and how to structure your collaboration with Claude.

A Systematic Framework for AI-Assisted Research

Before diving into specific techniques, let's establish a clear methodology connecting research principles to Claude collaboration:

Phase 1: Research Design

- Define clear research questions and objectives
- Identify required information types and sources
- Plan systematic search strategies
- Establish evaluation criteria for sources

Phase 2: Information Gathering

- Execute searches using multiple source types
- Apply verification principles consistently
- Document sources and findings systematically
- Maintain organized research workflow

Phase 3: Analysis and Synthesis

- Evaluate source credibility using established frameworks
- Identify patterns and relationships across sources
- Reconcile conflicting information and perspectives
- Generate insights through systematic comparison

Phase 4: Documentation and Application

- Create comprehensive summaries and reports
- Draw actionable conclusions from findings
- Acknowledge limitations and areas for further research
- Properly attribute sources and disclose AI assistance

What Is Research Methodology (and Why Should You Care)?

Research methodology isn't just a fancy term academics throw around—it's your systematic approach to gathering and analyzing information. Using research methodology vs. not to answer a question is like following a GPS route vs. wandering randomly through a city. Both might eventually get you somewhere, but only one reliably gets you where you need to go.

Good research methodology includes

- Clear objectives (what are you trying to find out?)
- Systematic search strategies (where and how to look)
- Critical evaluation (is this source reliable?)
- Organized synthesis (how do pieces fit together?)
- Actionable conclusions (what does it all mean?)

By systematically applying these steps with Claude's assistance, you can transform from someone who "looks stuff up" into someone who conducts actual research. Success requires deliberate implementation, not automatic improvement.

The Art of Analysis: More Than Just Reading

Before we dive into techniques, let's clarify what **analysis** really means. Analysis is the process of examining information in detail to understand it better or draw conclusions—it's the difference between reading a menu and understanding why the chef paired those flavors.

When you analyze with Claude, you're not just collecting facts. You're

- Identifying patterns and relationships
- Evaluating strengths and weaknesses
- Understanding context and implications
- Drawing meaningful conclusions
- Creating new insights from existing information

Think of Claude as an analytical collaborator that can potentially help you see connections based on pattern recognition in its training data, not novel analytical insight. The AI serves as a research partner that helps process information systematically, but human judgment remains essential for interpreting significance and drawing valid conclusions.

Starting Your Research Journey: The Right First Steps

Step 1: Define Your Research Question

Vague questions get vague answers. Before you start, get crystal clear on what you're trying to find out.

> **Bad Research Question**: "Tell me about electric cars"
>
> **Good Research Question**: "What are the main barriers to electric vehicle adoption in rural areas, and which solutions show the most promise?"
>
> **How Claude Helps**: "Claude, I need to research electric vehicles, but my topic feels too broad. Can you help me narrow it down to a specific, researchable question?"

Step 2: Create Your Research Plan

Don't just dive in—plan your approach. Claude can help you map out a research strategy:

> **You**: "I need to research the impact of remote work on productivity. Help me create a research plan."

Claude helps you identify

- Key aspects to investigate (individual productivity, team collaboration, work-life balance)
- Types of sources needed (academic studies, industry reports, case studies)
- Potential biases to watch for (tech company propaganda vs. traditional office advocates)
- Timeline and milestones for your research

Step 3: Cast Your Net (Strategically)

Now you're ready to gather information, but not like a hoarder at a garage sale. You want quality, not quantity.

Source Verification: Your BS Detector Upgrade

In our era of information overload, **source verification**—checking the reliability and accuracy of information sources—isn't optional. It's survival. Claude acts as your fact-checking partner, helping you separate gold from garbage.

The CRAAP Test (Yes, Really):

Librarians love this acronym for evaluating sources:

> **Currency**: When was this published? Is it still relevant?
> **Relevance**: Does this actually relate to your research question?
> **Authority**: Who wrote this? What are their credentials? **Accuracy**:

Can you verify this information elsewhere? **Purpose**: Why was this created? To inform, sell, or persuade?

Claude in Action: "Claude, I found this article claiming remote work increases productivity by 47%. Can you help me evaluate its credibility using the CRAAP test?"

Red Flags to Watch For

Claude can help you spot warning signs, though recognizing these patterns requires your guidance and verification:

- Emotional language pretending to be objective
- Cherry-picked statistics without context
- Missing citations or vague sources ("studies show...")
- Correlation presented as causation
- Outdated information presented as current

The Three-Source Rule

For any critical claim, find at least three independent sources. Claude can help interpret or reconcile differences between sources but cannot independently verify claims:

"Claude, three different sources claim different productivity increases from remote work: 13%, 22%, and 39%. How do I reconcile these differences?"

Claude might point out

- Different methodologies (self-reported vs. measured)
- Different definitions of "productivity"
- Different sample populations
- Different time periods studied

The Power of Synthesis: Creating New Understanding

Here's where the magic happens. **Synthesis** means combining different pieces of information to create new understanding—like a DJ mixing tracks to create something entirely new.

The Synthesis Process

This process involves collaborative work with Claude, where you provide guidance and Claude assists with analysis:

> **Step 1: Gather Your Pieces** Collect diverse perspectives on your topic. Don't just look for sources that agree with each other.
>
> **Step 2: Identify Patterns** *Claude Collaboration: "Help me identify themes and patterns across these sources. Where do experts agree? Where do they diverge?"*
>
> **Step 3: Connect the Dots** *Claude Prompt: "How do different pieces of information relate? What story emerges when you put them together?"*
>
> **Step 4: Generate Insights** *Claude Question: "What new understanding emerges that wasn't in any single source?"*

Synthesis in Action

Let's say you're researching productivity tools for your team:

> **You**: "Claude, I've researched 5 different project management tools. Here's what I found: [paste summaries]. Help me synthesize this into actionable insights."

Claude can help you see

- Common features across all tools (baseline requirements)
- Unique differentiators that matter for your use case

- Price-to-value relationships
- Hidden costs or limitations not obvious from marketing
- Which tool best fits your specific needs?

The synthesis creates understanding that no single review could provide.

Summarization: The Art of Distillation

Summarization—condensing longer content into shorter, key-point versions—is like making espresso from coffee. You're concentrating the essence while removing the fluff.

The Three Levels of Summarization

Level 1: Executive Summary (1-2 paragraphs) The absolute essence. What would a busy CEO need to know?

Level 2: Key Points Summary (1-2 pages) Main arguments, supporting evidence, critical data.

Level 3: Comprehensive Summary (2-5 pages) Detailed coverage while still significantly shorter than the original.

Summarization Techniques with Claude

The Progressive Summary:

1. "Summarize this article in one sentence"
2. "Now expand to a paragraph hitting the main points"
3. "Now create a bullet-point summary of key arguments and evidence"

The Perspective Summary: "Summarize this research paper as if explaining to

- A colleague in my field
- My manager who needs actionable insights
- A complete beginner"

The Critical Summary: "Summarize this article, highlighting

- Main claims
- Supporting evidence
- Potential weaknesses or biases
- Practical applications"

Real-World Research Scenarios

Let's see these principles in action across different research challenges:

Scenario 1: Market Research for a Startup

The Challenge: You need to understand the competitive landscape for your app idea.

The Systematic Approach:

1. Define specific research questions with Claude's help
2. Identify key sources (industry reports, competitor analyses, user reviews)
3. Verify credibility of market data using established frameworks
4. Synthesize findings into a competitive analysis through collaborative discussion
5. Create an actionable strategy document with proper attribution

Claude Prompt: "I'm researching the meditation app market. Help me create a research plan that covers market size, key players, user pain points, and opportunities for differentiation."

Scenario 2: Literature Review for Academic Work

The Challenge: Writing a literature review on climate change impacts on agriculture.

Note This focuses on the literature review process. Full academic research involves methodology design, data analysis, and peer review beyond the scope of this chapter.

The Approach:

1. Narrow topic to specific region and crops
2. Find peer-reviewed sources through systematic database searches
3. Verify scientific credibility using academic standards
4. Synthesize conflicting studies with Claude's analytical assistance
5. Create a comprehensive literature review with proper academic attribution

Important: Many institutions require explicit disclosure of AI assistance in academic work. Check your institutional policies before beginning.

Claude Prompt: "I'm researching climate change impacts on wheat production in the Midwest. Help me evaluate whether this source from [journal] is credible and how it fits with other research I've found."

Scenario 3: Personal Decision Research

The Challenge: Choosing the best neighborhood to move to.

The Approach:

1. Define criteria that matter (schools, commute, safety, amenities)
2. Gather data from multiple source types
3. Verify statistics and claims using triangulation
4. Synthesize findings into comparison matrix through structured analysis
5. Create summary findings for family discussion

Claude Prompt: "I'm researching neighborhoods in Seattle. Help me create a framework for comparing them across multiple factors and identify what sources would be most reliable for each type of information."

Systematic Research Techniques

Ready to level up? These systematic research practices will make you more effective. Note: These are established research practices enhanced by AI collaboration, not advanced methodologies unique to Claude.

The Reverse Research Method

Start with conclusions and work backward: "Claude, if someone claims that 'remote work reduces productivity,' what evidence would they need to support this? What counter-evidence should I look for?"

The Devil's Advocate Approach

For every source you find supporting your hypothesis, "Claude, play devil's advocate. What would critics say about this research? What contradicting evidence might exist?"

The Meta-Analysis Method

Instead of individual sources, look for patterns: "Claude, I've found 10 studies on this topic. Help me identify patterns in their methodologies, findings, and limitations."

The Timeline Technique

Track how understanding has evolved: "Claude, help me trace how expert opinion on [topic] has changed over the past decade. What drove these shifts?"

Building Your Research Workflow

Success isn't just about individual research projects—it's about building systems that make every research task easier.

The Research Documentation Framework

This framework serves as both a conversation structure with Claude and a documentation system for your research process:

Research Question Documentation

- Primary research question
- Sub-questions and scope limitations
- Success criteria for answering

Source Documentation

- Source identification and credibility assessment
- Key findings and relevance notes
- Bias identification and reliability evaluation

Analysis Documentation

- Pattern identification across sources
- Synthesis insights and new connections
- Contradictions and unresolved questions

Conclusion Documentation

- Summary and actionable conclusions
- Limitations and areas requiring further research
- Proper attribution and source acknowledgment

Claude Workflow Integration

Integrate Claude assistance into your personal research workflow:

Phase 1 Claude Prompts:

- "Help me refine this research question"
- "What types of sources should I prioritize?"
- "What potential biases should I watch for?"

Phase 2 Claude Prompts:

- "Evaluate this source using the CRAAP test"
- "Help me understand this contradictory information"
- "What questions does this source raise?"

Phase 3 Claude Prompts:

- "Help me synthesize these findings"
- "What patterns do you see across these sources?"
- "What am I missing or overlooking?"

Phase 4 Claude Prompts:

- "Help me create a clear summary"
- "What are the practical implications?"
- "How should I present this to [specific audience]?"

Common Research Pitfalls and How to Avoid Them:

Even with Claude's help, watch out for these traps:

Pitfall #1: Confirmation Bias

The Problem: Only finding sources that support what you already believe.

The Solution: Ask Claude to find opposing viewpoints: "What are the strongest arguments against my current conclusion?"

Pitfall #2: Source Overwhelm

The Problem: Collecting so many sources you can't process them.

The Solution: Focus on methodological appropriateness and relevance rather than volume. Have Claude help you identify the most methodologically sound and relevant sources first.

Pitfall #3: Analysis Paralysis

The Problem: Getting stuck in research mode without moving to conclusions.

The Solution: Set research deadlines. Ask Claude, "Based on what I've found so far, what preliminary conclusions can I draw?"

Pitfall #4: Shallow Synthesis

The Problem: Just summarizing without creating new insights.

The Solution: Push deeper: "Claude, what patterns or connections am I missing? What would an expert notice that I haven't?"

Pitfall #5: AI Over-reliance

The Problem: Accepting Claude's analysis without verification.

The Solution: Always cross-reference AI insights with original sources. Remember that Claude's pattern recognition is based on training data, not independent analysis.

Research Ethics and Integrity:

With great research power comes great responsibility:

Always Cite Sources

Claude helps track where information comes from, but you must credit original sources properly.

Acknowledge AI Assistance

Many professional and academic contexts require disclosure when AI tools assist with research. Check institutional policies and industry standards.

Acknowledge Limitations

No research is perfect. Be honest about what you don't know and where your methodology has constraints.

Avoid Misrepresentation

Use Claude to help rephrase and synthesize, not to misrepresent sources or create false attributions.

Respect Intellectual Property

Some information isn't meant to be freely shared. Respect paywalls, proprietary data, and copyright restrictions.

The Research Transformation Checklist:

Ready to revolutionize how you research? Here's your action plan:

Week 1: Foundation Building

- ☐ Practice defining clear research questions using the framework
- ☐ Master the CRAAP test for source evaluation
- ☐ Create your research documentation template

Week 2: Skill Development

- ☐ Complete one research project using the full systematic methodology
- ☐ Practice synthesis with 3-5 sources with Claude's assistance
- ☐ Create summaries at all three levels

Week 3: Advanced Applications

- ☐ Try reverse research and devil's advocate methods
- ☐ Build your source evaluation criteria for your field
- ☐ Experiment with meta-analysis approaches

Week 4: System Implementation

- ☐ Refine your research workflow with Claude integration
- ☐ Practice sharing findings using different formats
- ☐ Reflect on what systematic approaches work best for your needs

The Bottom Line

Research with Claude isn't about shortcuts—it's about being smarter, more thorough, and more insightful through systematic collaboration. You're not replacing your critical thinking; you're amplifying it with an AI partner who never gets tired of digging deeper while maintaining essential human oversight and verification.

Whether you're a student tackling assignments, a professional making strategic decisions, or just someone who loves learning, these systematic research skills will serve you for life. The internet has more information than any human could process. Claude helps you navigate it systematically while you maintain responsibility for verification and interpretation.

Remember: Good research isn't about finding the answer you want—it's about finding the truth, even when it's complicated, contradictory, or challenges your assumptions. Claude can help you process information systematically, but you're responsible for ensuring the analysis is sound and the conclusions are valid.

In the next chapter, we'll put your new research skills to work in a specific domain: coding and technical problem-solving. Even if you've never written a line of code, you're about to discover how Claude can make you dangerously effective with technology.

Ready to dive deep and emerge with insights? Your research transformation starts now.

CHAPTER 7

Coding and Programming with Claude

In This Chapter

- Discovering you can code (yes, you—the person who thinks they can't)
- Understanding programming languages without a computer science degree
- Learning to debug like a detective, not a psychic
- Mastering the art of code review with your AI mentor
- Building actual, working programs that do cool stuff

"I can't code." If I had a dollar for every time I've heard that, I could buy a small island and name it "Yes You Can Land." Here's the truth bomb: coding isn't about being a math genius or thinking in binary. It's about telling a computer what to do, step by step, in a language it understands. And with Claude as your patient, never-judgmental coding buddy, you're about to discover that programming is less like rocket science and more like following a really specific recipe.

Important Note This chapter focuses on fundamental programming concepts and basic projects. While Claude can be an excellent coding assistant, all code should be tested thoroughly, and production applications require additional security and error handling beyond these educational examples.

R. Dickey, *Mastering Claude AI*, https://doi.org/10.1007/979-8-8688-2001-4_7

A Framework for AI-Assisted Programming

Before diving into code, let's establish a systematic approach to learning programming with Claude's assistance:

Phase 1: Concept Understanding

- Ask Claude to explain programming concepts with analogies
- Request step-by-step breakdowns of complex ideas
- Get clarification on terminology and syntax rules

Phase 2: Guided Implementation

- Work with Claude to plan your code structure
- Get help writing functions and logic step-by-step
- Use Claude to explain what each part of your code does

Phase 3: Review and Improvement

- Ask Claude to review your code for errors and improvements
- Get suggestions for better practices and optimization
- Learn debugging strategies through collaborative problem-solving

Phase 4: Independent Practice

- Apply learned concepts to new problems
- Use Claude as a consultant for specific questions
- Build confidence through incremental challenges

This framework ensures you're learning transferable programming skills rather than just copying code.

The Great Coding Myth (and Why It's Complete Nonsense)

Let's destroy some myths right off the bat:

1. **Myth #1**: "You need to be good at math"

   Reality: Most basic coding is about logic, not calculus, though advanced domains like graphics, AI, and cryptography require substantial math. If you can follow a recipe or give driving directions, you can handle most programming logic.

2. **Myth #2**: "You have to start young"

   Reality: Some of the best programmers started in their 40s, 50s, or later. Your brain doesn't have an expiration date for learning.

3. **Myth #3**: "One mistake will break everything"

   Reality: That's what debugging is for! Every programmer spends more time fixing code than writing it.

4. **Myth #4**: "Real programmers memorize everything"

   Reality: Real programmers Google basic syntax daily and have Claude on speed dial.

What Is an Algorithm? (Spoiler: You Already Use Them)

Before we write a single line of code, let's understand what an **algorithm** really is. An algorithm is simply a step-by-step procedure or formula for solving a problem or completing a task. That's it. No magic, no mystery.

You use algorithms every day:

- **Making Coffee**: Boil water ➤ Add grounds ➤ Pour water ➤ Wait ➤ Drink
- **Getting Dressed**: Underwear ➤ Pants ➤ Shirt ➤ Shoes (usually in that order)
- **Making a Sandwich**: Bread ➤ Spread ➤ Filling ➤ Bread ➤ Cut ➤ Eat

Programming is just writing these instructions in a language computers understand. Instead of "make coffee," you might write more specific steps the computer can follow.

Your First Algorithm with Claude

Let's create an algorithm together using our framework approach:

> **You**: "Claude, help me write an algorithm for making the perfect grilled cheese sandwich. Break it down step-by-step and explain why each step matters."

Claude might respond with

1. Gather ingredients (bread, cheese, butter)—*Preparation prevents problems*
2. Heat pan to medium heat—*Too high burns the bread before the cheese melts*
3. Butter one side of each bread slice—*Creates the golden, crispy exterior*
4. Place the first slice butter-side down in the pan—*Buttered side touches the heat source*
5. Add cheese on top—*Even distribution ensures consistent melting*
6. Place the second slice butter-side up—*Prepares for the flip*
7. Cook until golden brown (about 3 minutes)—*Visual and time cues for doneness*
8. Flip sandwich—*Cook the other buttered side*
9. Cook the other side until golden brown—*Complete the process*
10. Remove from heat and let cool for 30 seconds—*Prevents mouth burns*
11. Cut diagonally (this is crucial!)—*Easier to eat and looks better*
12. Serve immediately—*Best texture and temperature*

See? You just created an algorithm. Coding is simply translating this structured thinking into a programming language.

Programming Languages: Choosing Your Weapon

A **programming language** is a formal language used to write computer programs—think of it as the specific dialect you use to talk to computers. Just like human languages, each has its own personality and best uses.

The Big Players (And When to Use Them):

Python: The Friendly Neighbor

- **Best For**: Beginners, data analysis, automation, AI/ML
- **Personality**: Like that teacher who makes everything clear
- **Claude Says**: "Start here if you're new to coding"

JavaScript: The Web Wizard

- **Best For**: Websites, web apps, interactive features
- **Personality**: The everywhere-at-once multitasker
- **Claude Says**: "Essential if you want to build anything web-related"

Java: The Corporate Professional

- **Best For**: Large applications, Android apps, enterprise software
- **Personality**: The suit-wearing, process-following executive
- **Claude Says**: "Great for learning structured programming"

SQL: The Data Whisperer

- **Best For**: Database queries, data manipulation
- **Personality**: The librarian who finds exactly what you need
- **Claude Says**: "A declarative programming language essential for data work"

How to Choose Your First Language

Ask Claude: "I want to learn programming to [your goal]. Which language should I start with?"

Implementation Framework for Language Selection:

Step 1: Define Your Goals

- "Automate boring tasks" ➤ Python
- "Build websites" ➤ JavaScript + HTML/CSS
- "Analyze data" ➤ Python or SQL
- "Make mobile apps" ➤ Swift (iOS) or Kotlin (Android)
- "Just learn to code" ➤ Python (it's the friendliest)

Step 2: Consider Your Learning Style

- Visual learner ➤ JavaScript (immediate visual feedback)
- Logic-focused ➤ Python (clean, readable syntax)
- Structure-oriented ➤ Java (enforces good practices)

Step 3: Evaluate Resources

- Available time for learning
- Access to practice environments
- Community support and tutorials

Understanding Syntax: The Grammar of Code

Syntax refers to the rules governing the structure of programming languages—it's like grammar for code. Just as "Dog the brown quick" doesn't make sense in English, computers are picky about code structure.

Syntax Basics Every Language Shares

Variables: Labeling Your Stuff

Think of variables as labeled boxes where you store information:

```
name = "Claude"          # Text storage
age = 2                  # Number storage
is_helpful = True        # Boolean (True/False) storage
```

Functions: Reusable Instructions

Functions are like recipes you can use repeatedly. Here's the recipe definition:

```
def make_greeting(name):          # The recipe: takes a name as input
    return f"Hello, {name}!"      # The process: creates greeting text
```

And here's making the dish (using the recipe):

```
greeting = make_greeting("World") # Use the recipe with "World"
print(greeting)                   # Outputs: "Hello, World!"
```

Conditions: Making Decisions

Code that makes choices based on conditions:

```
if temperature > 80:                  # If temperature is above 80...
    print("It's hot! Drink water!")   # ...say "It's hot! Drink water!"
else:                                 # But if it isn't above 80...
    print("Nice weather!")            # ...say "Nice weather!"
```

Loops: Repetition Without Repetitive Strain Injury

Do something multiple times without writing it multiple times:

```
for i in range(5):                                       # Repeat 5 times
    print("I will not throw paper airplanes in class")  # Print this
                                                           each time
```

When Syntax Attacks: Common Mistakes

Claude can help you avoid these classic blunders by explaining what went wrong:

- Missing colons (Python's favorite gotcha)
- Forgetting parentheses or brackets
- Mixing up = (assignment) and == (comparison)
- Incorrect indentation (Python is picky about this)
- Forgetting semicolons (JavaScript's nemesis)

Implementation Tip Ask Claude, "What are the most common syntax errors beginners make in [language], and how can I avoid them?"

Essential Programming Concepts You Need to Know

Before jumping into projects, let's build your foundation with core concepts:

Data Types: What Kind of Information You're Storing

Strings: Text information

```
name = "Alice"
message = "Hello, how are you?"
```

Integers: Whole numbers

```
age = 25
count = 100
```

Floats: Decimal numbers

```
price = 19.99
temperature = 98.6
```

Lists: Collections of items

```
colors = ["red", "green", "blue"]
numbers = [1, 2, 3, 4, 5]
```

Dictionaries: Key-value pairs (like a phone book)

```
person = {
    "name": "Alice",
    "age": 25,
    "city": "Portland"
}
```

Variable Scope: Where Your Variables Live

Variables have different "lifespans" depending on where they're created:

```
global_variable = "Everyone can see me"

def my_function():
    local_variable = "Only this function can see me"
    print(global_variable)     # This works - can see global
    print(local_variable)      # This works - local to function

print(global_variable)         # This works
print(local_variable)          # This fails - variable doesn't exist here
```

Function Parameters and Returns: Communication Between Code Sections

Functions can receive information (parameters) and send information back (return values):

```
def calculate_area(length, width):       # Function receives two parameters
    area = length * width                # Calculate using the parameters
    return area                          # Send the result back

# Using the function
room_area = calculate_area(10, 12)       # Send 10 and 12 to the function
print(f"Room area: {room_area}")         # Use the returned result
```

Input Validation: Checking What Users Give You

Always verify that user input is what you expect:

```
def get_positive_number(prompt):
    while True:                              # Keep asking until valid
        try:                                 # Try to convert input
            number = float(input(prompt))    # Get input and convert
                                               to number
            if number > 0:                   # Check if positive
                return number                # Return valid number
```

```
        else:
            print("Please enter a positive number.")
    except ValueError:                    # If conversion fails
        print("Please enter a valid number.")

# Usage
age = get_positive_number("Enter your age: ")
```

Your First Real Program (It's Easier Than You Think)

Let's build something real using our systematic framework. We'll create a password strength checker with proper educational practices.

Step 1: Plan with Claude

You: "Claude, I want to create a simple password strength checker in Python. Help me plan what it should do, and explain the programming concepts I'll learn."

Claude helps you outline:

1. Ask the user for a password (user input, string handling)
2. Check password length (string operations, conditionals)
3. Check for uppercase letters (loops, string methods)
4. Check for lowercase letters (Boolean logic)
5. Check for numbers (pattern recognition)
6. Check for special characters (string operations)
7. Give feedback on strength (functions, return values)
8. Handle errors gracefully (exception handling)

Step 2: Write the Code with Educational Focus

You: "Great! Now help me write the code for this, explaining each programming concept as we use it."

Claude provides code with comprehensive explanations:

```
# Password Strength Checker - Educational Example
# This program demonstrates: functions, loops, conditionals, lists,
# string methods, and basic security awareness

def check_password_strength(password):
    """
    Check password strength and return feedback.

    This function demonstrates:
    - Function definition with parameters and return values
    - Local variables and scope
    - Conditional statements (if/elif/else)
    - List operations (append)
    - String methods (isupper, islower, isdigit)
    - Generator expressions with any()
    - Tuple returns (multiple values)
    """

    # Initialize our score (integer variable)
    score = 0
    feedback = []  # List to store suggestions

    # Check length (string length, conditional logic)
    if len(password) >= 8:
        score += 1
    else:
        feedback.append("Use at least 8 characters")

    # Check for uppercase letters (any() function with generator
      expression)
    if any(char.isupper() for char in password):
        score += 1
    else:
        feedback.append("Add uppercase letters")
```

```
    # Check for lowercase letters (iteration and boolean logic)
    if any(char.islower() for char in password):
        score += 1
    else:
        feedback.append("Add lowercase letters")

    # Check for numbers (string method for digit detection)
    if any(char.isdigit() for char in password):
        score += 1
    else:
        feedback.append("Add numbers")

    # Check for special characters (membership testing)
    special_chars = "!@#$%^&*()_+-=[]{}|;:,.<>?"
    if any(char in special_chars for char in password):
        score += 1
    else:
        feedback.append("Add special characters")

    # Determine strength based on score (conditional chain)
    if score == 5:
        strength = "Very Strong! 🎉"
    elif score == 4:
        strength = "Strong 💪"
    elif score == 3:
        strength = "Moderate 😐"
    elif score == 2:
        strength = "Weak 😟"
    else:
        strength = "Very Weak 😱"

    # Return multiple values as a tuple
    return strength, feedback

def get_secure_password_input():
    """
    Get password input with educational security note.
```

Educational Note In real applications, use the getpass module to hide password input from the terminal and command history.

```
    Example for production:
     import getpass
     password = getpass.getpass("Enter password: ")
     """
     print("Educational Note: This is for learning only.")
     print("Real apps should hide password input for security.")
     return input("Enter a password to check: ")

def main():
    """
    Main program function demonstrating program flow and error handling.

    This demonstrates:
    - Function organization
    - Error handling with try/except
    - User interaction flow
    - String formatting
    """

    print("Password Strength Checker")
    print("-" * 25)

    try:
        # Get password from user with security awareness
        password = get_secure_password_input()

        # Check the password (function call with return value capture)
        strength, suggestions = check_password_strength(password)

        # Display results (string formatting and conditional output)
        print(f"\nPassword Strength: {strength}")

        if suggestions:  # List evaluation (empty list is False)
            print("\nSuggestions to improve:")
```

```
            for suggestion in suggestions:  # List iteration
                print(f" • {suggestion}")
        else:
            print("\nYour password is excellent!")

    except KeyboardInterrupt:  # Handle Ctrl+C gracefully
        print("\n\nProgram interrupted by user.")
    except Exception as e:  # Catch unexpected errors
        print(f"\nAn error occurred: {e}")
        print("Please try again.")

# Run the program (this only runs when script is executed directly)
if __name__ == "__main__":
    main()
```

Step 3: Understand What You Built

With Claude's help, you just created a real, working program that demonstrates:

Core Programming Concepts:

- Variables and data types (strings, integers, Booleans, lists)
- Functions with parameters and return values
- Conditional statements (if/elif/else chains)
- Loops and iteration (for loops with generator expressions)
- Error handling (try/except blocks)
- String operations and methods
- List operations and Boolean evaluation

Security Awareness:

- Recognition of password input security issues
- Proper commenting and documentation
- Error handling for robustness

Best Practices:

- Function organization and single responsibility
- Clear variable names and comments
- User-friendly output formatting
- Graceful error handling

Not bad for someone who "can't code," right?

Development Environment: Your Coding Command Center

An **IDE (Integrated Development Environment)** is software that provides comprehensive tools for coding—think of it as Microsoft Word for programmers, but way cooler. You don't need one to start (you can code in Notepad), but they make life much easier.

Recommended Development Environments for Beginners

Online Environments (No Installation Needed):

Replit: The Browser-Based Buddy

- No installation needed
- Built-in collaboration features
- Automatic saving and sharing
- Perfect safe sandbox for learning
- Claude says "Ideal for beginners and educational projects"

CodePen: The Web Playground

- Great for HTML/CSS/JavaScript
- Instant visual feedback
- Large community of examples

Local Development Environments:

VS Code: The Swiss Army Knife

- Free, powerful, works for everything
- Extensions for every language
- Built-in terminal and debugging
- Claude recommends "Start here for serious development"

PyCharm Community: The Python Specialist

- Specifically designed for Python
- Excellent error detection and hints
- Great debugging tools
- Claude says "Excellent for Python-focused learning"

IDE Features That Save Your Sanity

Syntax Highlighting: Different colors for different code parts (like highlighting important text)

Auto-Completion: The IDE suggests what you might type next (like your phone's predictive text, but actually helpful)

Error Detection: Red squiggly lines under mistakes (like spell check for code)

Debugging Tools: Step through code line by line to find problems

Integrated Terminal: Run your code without leaving the IDE

Implementation Tip Ask Claude, "What IDE features should I learn first in [your chosen IDE], and how do they help with [your programming language]?"

Debugging: Becoming a Code Detective

Debugging—finding and fixing errors in computer code—is where you'll spend most of your coding time. But don't worry, it's actually the fun part (seriously!). It's like being a detective, but instead of solving crimes, you're solving why your program thinks 2 + 2 = 22.

Types of Bugs You'll Meet

Syntax Errors: The Grammar Police

Your code breaks the language rules:

```
print("Hello World"    # Missing closing parenthesis
```

Logic Errors: The Sneaky Ones

Your code runs but does the wrong thing:

```
# Trying to calculate average
total = 100
count = 3
average = total + count    # Should be division, not addition!
```

Runtime Errors: The Surprise Party Crashers

Your code works until it doesn't:

```
numbers = [1, 2, 3]
print(numbers[10])         # There is no item at index 10! (Python uses
                             0-based indexing:
                           # index 0=1st item, index 1=2nd item, index
                             2=3rd item)
```

Debugging Strategies with Claude

The Rubber Duck Method (Explaining to Find Your Own Answers)

Here's a programmer's secret: explaining your problem out loud often helps you solve it yourself. Traditional programmers explain code to a rubber duck on their desk. You've got something better—Claude!

Start explaining your code line by line: "Claude, I'm trying to calculate the average but getting weird results. Here's my code..." Often, just describing the problem helps you spot the error yourself. And unlike a rubber duck, if you don't see it, Claude can actually help!

Systematic Code Investigation

When code fails, narrow down the problem systematically:

> **Prompt for Claude**: "Claude, my program crashes somewhere. Help me add print statements to figure out where the problem occurs, and explain why this debugging approach works."

The Fresh Eyes Approach

Sometimes you're too close to see the obvious:

> **Prompt for Claude**: "Claude, this code should work but doesn't. What am I missing? Please explain any errors you spot and why they cause problems."

The Error Message Translation

Found an error message you don't understand?

> **Prompt for Claude**: "Claude, I'm getting this error: [paste error]. What does it mean in plain English, and what are the most common causes?"

Debugging in Action

Let's debug a real problem with Claude's assistance:

> **You**: "Claude, my code is supposed to reverse a word, but it's not working:"

```
def reverse_word(word):
    reversed = ""
    for i in range(len(word)):
        reversed = word[i] + reversed
    return reverse  # Bug is here!

print(reverse_word("hello"))
```

Claude Might Help Identify: "You have a typo! The function returns 'reverse' but your variable is named 'reversed.' Python doesn't know what 'reverse' is."

One character fixed, problem solved. Most bugs are like this—small typos with big consequences.

Implementation Framework for Debugging:

1. **Read the error message carefully** (Claude can translate if confusing)
2. **Identify the line number** where the error occurs
3. **Check for common syntax mistakes** (missing punctuation, typos)
4. **Verify variable names** match throughout your code
5. **Test with simple inputs** to isolate the problem
6. **Use print statements** to see what values variables actually contain

Code Review: Learning from Your Code

Code review is the process of examining code for errors, improvements, or optimization opportunities. It's like having someone proofread your essay, but for code. Claude is your always-available code reviewer.

What to Review:

Functionality: Does it do what it's supposed to?

- Test with different inputs
- Check edge cases (what if the list is empty?)
- Verify expected outputs

Readability: Can someone else understand it?

- Clear variable names (not x, y, z)
- Comments explaining complex parts
- Logical organization

Efficiency: Could it be faster or use less memory?

- Avoid unnecessary loops
- Use built-in functions when available
- Don't repeat yourself (DRY principle)

Best Practices: Does it follow conventions?

- Consistent naming style
- Proper indentation
- Error handling where appropriate

Code Review Prompts for Claude

Basic Review: "Claude, review this code for any obvious issues or improvements. Explain why each suggestion would make the code better."

Specific Review: "Claude, is there a more efficient way to write this function? Show me the improvement and explain the benefits."

Learning Review: "Claude, what programming concepts does this code demonstrate? How could I make it more professional?"

Security Review: "Claude, are there any security issues with this educational code? What would I need to consider for a production version?"

Real-World Programming Projects:

Let's apply everything with practical projects using our systematic framework:

Project 1: Enhanced To-Do List Manager

Goal: Create a program to manage daily tasks with file persistence

Skills Practiced:

- User input and validation
- Data storage (lists and dictionaries)
- File handling (reading/writing)
- Menu-driven interface
- Error handling

Programming Concepts Covered:

- Data structures (lists, dictionaries)
- File I/O operations
- Exception handling
- Function organization
- User interface design

Start With: "Claude, help me create a command-line to-do list manager in Python that saves tasks to a file. Break this into steps and explain what programming concepts each step teaches."

Project 2: Simple Calculator with History

Goal: Build a calculator that remembers previous calculations

Skills Practiced:

- Mathematical operations
- Input validation
- Data persistence
- Menu systems
- Error handling for mathematical edge cases

Programming Concepts Covered:

- Arithmetic operations and operator precedence
- Exception handling for division by zero
- List management for history
- String parsing and conversion
- Loop structures for continuous operation

Start With: "Claude, I want to build a calculator that keeps track of calculation history. Guide me through the design and explain how this teaches fundamental programming concepts."

Project 3: Personal Expense Tracker

Goal: Track income and expenses with basic reporting

Skills Practiced:

- Data organization and categorization
- Basic statistical calculations
- Date handling
- File operations
- Simple reporting

Programming Concepts Covered:

- Dictionary data structures for categorization
- Date and time operations
- List comprehensions for data filtering
- Basic statistical functions
- Formatted output and reporting

Start With: "Claude, help me build a simple expense tracker that categorizes spending and provides basic reports. Show me how this project teaches data organization and analysis concepts."

Implementation Timeline (Realistic 6-Week Plan)

Weeks 1-2: Foundation Building

- Master basic syntax and data types
- Complete the password checker project
- Practice with simple functions and conditionals
- Learn basic debugging techniques

Weeks 3-4: Intermediate Concepts

- Work with lists and dictionaries
- Practice file I/O operations
- Build the to-do list manager
- Learn error handling

Weeks 5-6: Project Development

- Choose between calculator or expense tracker
- Plan the project with Claude's help
- Implement incrementally with testing
- Practice code review and improvement

Common Coding Fears and How to Overcome Them:

Fear #1: "I'll Break Something"

Reality: In educational environments like Replit or basic Python scripts, you're working in a safe environment. The worst that happens is your program stops running, and you restart it.

Claude Helps: "Is there any way this educational code could damage my computer?" (Spoiler: Basic learning code in safe environments cannot)

Implementation Tip Always use educational platforms or sandboxed environments for learning.

Fear #2: "I'm Too Slow"

Reality: Every expert was once slow. Speed comes with pattern recognition, which develops through practice.

Claude Helps: "This simple task took me an hour. Is that normal for beginners? What can I do to improve my learning efficiency?"

Fear #3: "Everyone Else Gets It"

Reality: Everyone Googles basic syntax. Everyone forgets semicolons. Everyone debugs more than they code. Programming forums exist because even experts need help.

Claude Helps: "I feel like I'm the only one struggling with [concept]. Can you explain it differently and share some encouragement about the learning process?"

Fear #4: "I'm Not Smart Enough"

Reality: Coding is a skill, not an IQ test. Like cooking or driving, it improves with practice and patience.

Claude Helps: "I don't understand [concept] even after reading about it. Can you explain it with a real-world analogy and give me a simple practice exercise?"

Your Coding Journey Checklist:

Ready to start coding? Here's your systematic path:

Week 1: Foundation

- ☐ Choose your first programming language based on goals
- ☐ Set up a development environment (Replit for simplicity)
- ☐ Write your first "Hello World" program
- ☐ Create the password checker with Claude's guidance
- ☐ Learn basic debugging with print statements

Week 2: Core Concepts

- ☐ Master variables and data types with practice exercises
- ☐ Practice with conditions (if/else) using real scenarios
- ☐ Understand basic loops with repetitive tasks
- ☐ Debug your first logical error with Claude's help
- ☐ Learn function basics and parameter passing

Weeks 3–4: Real Projects

- ☐ Plan a simple calculator with Claude
- ☐ Implement basic arithmetic operations
- ☐ Add input validation and error handling
- ☐ Start your to-do list project systematically
- ☐ Practice code review sessions with Claude

Weeks 5–6: Confidence Building

- ☐ Complete one project from start to finish
- ☐ Test your project thoroughly with various inputs

☐ Document your code with comments and explanations

☐ Share your progress and get feedback

☐ Plan your next learning goals with specific objectives

Security and Best Practices Awareness:

Educational vs. Production Code

Important Distinction: The code examples in this chapter are designed for learning fundamental concepts. Production applications require additional considerations:

Security Considerations:

- Password input should use secure methods (getpass module)
- User input needs comprehensive validation
- File operations require permission and error checking
- Network code needs encryption and authentication

Professional Development Practices:

- Comprehensive error handling and logging
- Code testing and validation
- Documentation and version control
- Performance optimization and scalability

Implementation Tip Always ask Claude, "What would I need to change in this educational code to make it suitable for real-world use?"

Building Good Habits Early

Code Organization:

- Use clear, descriptive variable names
- Write functions that do one thing well

- Comment on complex logic and decisions
- Organize code into logical sections

Error Handling:

- Always validate user input
- Handle expected errors gracefully
- Provide helpful error messages
- Test edge cases and unusual inputs

Security Mindset:

- Never trust user input without validation
- Understand what your code can access
- Keep sensitive information secure
- Learn about common security vulnerabilities

The Bottom Line

Coding isn't about becoming the next tech billionaire (though that's a nice side effect). It's about solving problems, automating boring stuff, and creating things that didn't exist before. With Claude as your patient teacher, you can learn at your own pace without judgment.

Remember: Every programmer you admire once wrote their first "Hello World" and immediately got an error. The difference between you and them isn't talent—it's that they kept going. And with Claude by your side, you will too.

Your code doesn't have to be perfect. It just has to work. And when it doesn't work (which will happen a lot), you now know how to fix it systematically. That's not failing—that's programming.

The key insights for successful AI-assisted programming:

> **Start with Understanding**: Always ask Claude to explain concepts before jumping into code. **Build Systematically**: Use the framework approach to connect each lesson to the next. **Practice Deliberately**: Apply concepts to real problems that interest you.

Debug Confidently: Treat errors as puzzles to solve, not failures to avoid. **Review Regularly**: Use Claude to help improve your code and learn better practices.

In the next chapter, we'll explore creative problem-solving and innovation with Claude. But for now, open that development environment and write some beautifully imperfect code. Your computer is waiting for instructions.

Ready to speak to a computer? Let's code!

CHAPTER 8

Creative Projects and Problem-Solving

In This Chapter

- Discovering your inner innovator (yes, even if you think you're "not creative")
- Mastering brainstorming techniques that actually work
- Understanding design thinking without needing a design degree
- Building creative collaborations between you and Claude
- Turning wild ideas into real-world solutions
- Strategic planning that doesn't feel like homework

"I'm not creative." Stop right there. If you've ever figured out how to untangle Christmas lights, invented a new sandwich combination, or found a clever excuse for being late, you're creative. Creativity isn't about painting masterpieces or composing symphonies—it's about solving problems in new ways. And with Claude as your **creative collaboration** partner, you're about to discover that **innovation** isn't reserved for Silicon Valley wizards or artistic geniuses. It's for anyone willing to ask, "What if?" and "Why not?"

R. Dickey, *Mastering Claude AI*, https://doi.org/10.1007/979-8-8688-2001-4_8

Important Note Claude provides pattern-based creative suggestions from its training data, not genuine creative understanding. It cannot truly evaluate artistic merit or provide authentic creative critique. While Claude can assist with creative processes, breakthrough innovation requires human insight, domain expertise, and real-world implementation that AI cannot provide.

A Framework for AI-Assisted Creative Collaboration

Before diving into creative techniques, let's establish a systematic approach to working creatively with Claude:

Phase 1: Creative Problem Definition

- Use Claude to help clarify and reframe creative challenges
- Generate multiple perspectives on the same problem
- Identify constraints that can actually enhance creativity
- Establish clear success criteria for creative outcomes

Phase 2: Ideation and Exploration

- Leverage Claude's pattern recognition for idea generation
- Use specific prompting techniques for diverse creative directions
- Combine human intuition with AI's broad knowledge synthesis
- Apply systematic brainstorming methodologies

Phase 3: Development and Refinement

- Use Claude to help structure and organize creative concepts
- Get assistance with research and background information
- Iterate through multiple versions with AI feedback
- Apply design thinking principles systematically

Phase 4: Implementation and Validation

- Utilize Claude for planning and project management support
- Get help with documentation and communication
- Use AI assistance for routine tasks while maintaining creative ownership
- Validate concepts through human judgment and real-world testing

This framework ensures you're building genuine creative capabilities while leveraging AI's strengths appropriately.

The Creative Myth (Time to Bust It Wide Open)

Let's destroy some creativity myths that have been holding you back:

1. **Myth #1: "Creativity is something you're born with"**
   Reality: Creativity is a skill, like cooking or driving. You get better with practice and systematic approaches.
2. **Myth #2: "Creative people are always having brilliant ideas"**
   Reality: Creative people have lots of terrible ideas. They just keep going until they find good ones.
3. **Myth #3: "You need inspiration to be creative"**
   Reality: Inspiration is nice, but a systematic process beats waiting for lightning to strike.
4. **Myth #4: "AI kills creativity"**
   Reality: AI can serve as a brainstorming partner, but it's based on pattern recognition, not genuine creative insight. The key is understanding how to collaborate effectively.

What Is Creativity, Really? (Spoiler: You Already Have It)

Creativity is simply the ability to see connections others miss and combine existing things in new ways. Every time you

- Find a new route to avoid traffic
- Repurpose a paper clip as a phone stand
- Combine leftovers into a surprisingly good meal
- Make up a bedtime story for a kid

You're being creative. It's not magic—it's problem-solving with style.

Human vs. AI Creative Contributions:

Understanding what each partner brings to creative collaboration:

Human Strengths:

- Intuition and emotional understanding
- Real-world experience and context
- Value judgment and artistic sensibility
- Problem identification and goal setting
- Cultural and social awareness

Claude's Capabilities:

- Pattern recognition across vast information sets
- Rapid generation of combinatorial possibilities
- Systematic exploration of idea spaces
- Research and information synthesis
- Consistent availability for iteration

Effective Collaboration: Combines human creativity and judgment with AI's processing and pattern recognition capabilities.

Brainstorming: Your Creative Superpower

Brainstorming—generating lots of ideas without judging them—is where creativity begins. Most people do it wrong by immediately shooting down ideas. With Claude, you have a judgment-free zone where every idea, no matter how wild, gets explored.

The Systematic Brainstorming Method

Rather than hoping for random inspiration, use this structured approach:

Step 1: Problem Framing with Claude

Prompt Template: "Claude, I need to [specific goal]. Help me reframe this problem from multiple angles. What are 5 different ways to think about this challenge?"

You: "Claude, I need to plan a memorable 10-year anniversary celebration, but I'm stuck. Help me reframe this problem from multiple angles."

Claude Might Respond with Different Framings:

- **Memory Celebration**: "How can we honor the journey you've taken together?"
- **Future Visioning**: "How can you launch the next decade of your relationship?"
- **Experience Design**: "What kind of experience would surprise and delight both of you?"
- **Story Creation**: "How can you create a new chapter in your love story?"
- **Value Expression**: "How can you express what matters most about your partnership?"

Step 2: Systematic Idea Generation

Prompt Template for Different Approaches:

Traditional Brainstorming: "Generate 10 ideas for [problem], ranging from conventional to completely unconventional. Don't filter for feasibility yet."

Constraint-Based Thinking: "Generate ideas for [problem] with these constraints: [budget/time/location limits]. How do limitations spark creativity?"

Opposite Thinking: "What would be the exact opposite of how people usually approach [problem]? Generate ideas based on doing the reverse."

Analogical Thinking: "How do [completely different fields/ industries] solve similar challenges? Generate ideas by borrowing from unexpected sources."

Advanced Brainstorming Techniques with Implementation

The "Yes, And" Method

Borrowed from improv comedy, this builds on ideas instead of blocking them:

Implementation with Claude:

You: "I want to start a small business but everything seems taken."

Claude: "Let's 'Yes, And' this. What's something you enjoy doing?"

You: "I like organizing things."

Claude: "Yes, and organizing is a superpower many people lack. What if you organized people's digital lives—photos, passwords, files?"

You: "Yes, and I could create systems they can maintain themselves..."

Claude: "Yes, and you could offer quarterly 'tune-ups' for recurring revenue..."

Systematic Prompting: "Help me build on this idea using 'Yes, And' thinking: [initial idea]. Generate 3 extensions, then help me build on the most promising one."

The Forced Connection Method

Force connections between unrelated things:

Implementation Prompt: "Connect [concept A] with [concept B] in 5 different ways. Then develop the most interesting connection into a concrete idea."

You: "Claude, connect 'pizza delivery' with 'library services'—go wild."

Claude: "Book delivery service with 30-minute guarantees? Late-night study snack service for libraries? 'Pizza party' reading programs? A mobile library that delivers books like pizza—you order online, they come to you?"

Design Thinking: Solving Problems Like a Pro

Design thinking sounds fancy, but it's just a structured way to solve problems by focusing on people's needs. You don't need a design degree—you just need empathy and curiosity.

The Five Stages with Claude Implementation

1. **Empathize: Understand the Real Problem**

   Implementation Framework:

   Before solving, understand who you're solving for:

   Prompt Template: "Help me understand [target audience] better. What are their real barriers to [desired outcome]? What motivates them? What frustrates them? Generate 5 specific user scenarios."

   You: "I want to create an app that helps people exercise more."

   Claude: "Let's empathize first. Who specifically are you helping? Busy parents? Seniors? Teens? What are their real barriers to exercise? Time? Motivation? Knowledge? Physical limitations? Social anxiety?"

2. **Define: Frame the Right Question**

   Implementation Framework:

   Turn vague problems into specific challenges:

Prompt Template: "Help me reframe this problem statement: [vague problem]. Create 3 'How might we...' questions that are specific, actionable, and human-centered."

You: "People don't exercise enough."

Claude: "Let's reframe with specific 'How might we' questions:

- How might we help busy parents find 10 minutes for movement that doesn't feel like exercise?
- How might we make physical activity social and fun for isolated seniors?
- How might we create exercise habits that work with unpredictable schedules?"

3. **Ideate: Generate Solutions**

   Implementation Framework:

   Brainstorm with purpose using the methods above, but now focused on your defined problem.

   Prompt Template: "For this specific problem: [defined challenge], generate 10 solutions using these approaches: [conventional, technology-enabled, social/community-based, behavior-change-focused, and completely unconventional]."

4. **Prototype: Make It Real (But Simple)**

   Implementation Framework:

   Start small and test fast:

   Prompt Template: "Help me create the simplest possible version of [idea] that I could test with 5 people this week. What's the minimum viable version that would give me useful feedback?"

   You: "The playground workout sounds good, but I can't code an app."

   Claude: "Start simpler: Create a PDF guide. Test with 5 parents. Get feedback. Then maybe a simple website. Then consider an app. Always start with the simplest version that could work."

5. **Test: Learn and Improve**

   Implementation Framework:

   Real feedback beats perfect planning:

   Prompt Template: "After testing [prototype] with [number] people, help me analyze this feedback: [feedback]. What patterns do you see? What should I prioritize changing?"

   Claude: "After your test, ask: Did parents actually use it? What stopped them? What surprised you? What would they pay for? Each answer improves your next version."

Creative Project Development: From "What If" to "Ta-Da!"

Let's walk through real creative projects using Claude as your systematic collaboration partner.

Project 1: Family Story Preservation

Goal: Document family memories in an engaging way

Implementation Framework:

Phase 1: Problem Definition Prompt: "I want to preserve my grandparents' stories but don't know where to start. Help me break this into specific, manageable components."

Claude Helps You Structure:

1. Story collection strategy (interview techniques, conversation starters)
2. Organization themes (chronological, topical, character-focused)
3. Format options (written memoir, audio recordings, photo story, cookbook with stories)
4. Family involvement approaches (collaborative vs. individual)

5. Preservation and sharing methods

Phase 2: Systematic Implementation Prompt: "Help me create a step-by-step plan for [chosen approach] that I can complete in [timeframe] with [available resources]."

Phase 3: Creative Enhancement Prompt: "What creative elements could make this family history more engaging? Suggest 5 approaches that go beyond traditional photo albums."

The Claude Advantage: Never runs out of follow-up questions, helps organize scattered memories, and suggests creative presentation formats—but you provide the family knowledge and emotional connections.

Project 2: The Side Hustle Developer

Goal: Turn a skill into income

Implementation Framework:

Phase 1: Skill Reframing Prompt: "I'm good at [skill] but that seems boring for a business. Help me reframe this skill in 5 different ways that could create value for specific audiences."

You: "I'm good at Excel but that seems boring for a side business."

Claude Reframes It: "Excel skills are problem-solving skills! You could:

- Create custom automation for small businesses
- Teach 'Excel for Real Life' workshops to specific groups
- Build industry-specific templates (restaurants, freelancers, nonprofits)
- Offer 'Excel Rescue' services for overwhelmed entrepreneurs
- Create Excel-based tools for specific problems (budget trackers, project planners)"

Phase 2: Market Validation Prompt: "For this idea: [specific reframed skill application], help me design 5 questions to validate market demand before I invest significant time."

Phase 3: MVP Development Prompt: "What's the absolute simplest version of [validated idea] I could create and test this month with a budget of $[amount]?"

The Claude Advantage: Sees opportunities you miss, connects skills to market needs, and helps position "ordinary" as valuable—but you bring domain expertise and market execution.

Project 3: Learning Through Creation

Goal: Learn something new in an engaging way

Implementation Framework:

Phase 1: Learning Goal Deconstruction Prompt: "I want to learn [subject] but traditional methods are boring. Help me design 5 creative learning approaches that involve making something tangible."

You: "I want to learn Spanish but apps are boring."

Claude Creates a Custom Approach: "Let's make Spanish learning creative:

- Document your daily life in three-sentence Spanish diary entries
- Translate and perform your favorite songs
- Create a Spanish-speaking alter ego with a backstory and social media presence
- Cook international recipes following Spanish-only instructions
- Write and illustrate a simple comic strip in Spanish
- Start a Spanish meme collection with cultural context notes"

Phase 2: Systematic Structure Prompt: "Create a 30-day challenge structure for [chosen learning approach] that builds skills progressively while staying engaging."

Phase 3: Community and Accountability Prompt: "How can I add social elements to this learning project to increase motivation and create accountability?"

The Claude Advantage: Personalizes learning to your interests, creates systematic progression, and suggests accountability methods—but you provide motivation and consistent practice.

Strategic Planning: Big Dreams, Real Steps

Strategic planning isn't just for corporations. It's how you turn "someday" dreams into "next Tuesday" actions. Claude excels at breaking overwhelming goals into manageable pieces, though all plans require human validation for feasibility, resource availability, and contextual appropriateness.

The Backward Design Method

Start with success and work backward:

> **Implementation Prompt:** "Help me work backward from this goal: [big goal]. Break it into yearly, monthly, weekly, and daily actions, noting resource requirements and potential obstacles at each level."
>
> **You:** "I want to write a novel but it feels impossible."
>
> **Claude Walks You Backward:** "Finished novel = 80,000 words Let's work backward realistically:
>
> - 250 words per weekday (1 page) = 1,250 words/week
> - 1,250 × 50 weeks = 62,500 words
> - Add weekends for editing = 70,000+ words in one year
> - Weekly goals: 5 pages written + 1 page edited
> - Daily goals: Write for 30 minutes before breakfast
> - Resource needs: Quiet space, simple word processor, story outline
> - Potential obstacles: Schedule changes, motivation dips, perfectionism"

The Stepping Stone Strategy

Break overwhelming changes into manageable steps:

> **Implementation Prompt:** "I want to transition from [current situation] to [desired outcome] but the gap feels huge. Create a 6-month stepping stone plan where each month builds on the previous one."
>
> **You:** "I want to change careers to UX design but have no experience."

Claude Creates Stepping Stones:
Month 1: Foundation Building

- Complete one comprehensive UX course (Coursera/Udemy)
- Read 2 industry-standard books
- Join 3 UX communities online
- Start following 10 UX professionals on LinkedIn

Month 2: Skill Development

- Complete first practice project (redesign existing app)
- Learn design tools (Figma basics)
- Attend 2 virtual UX meetups
- Start UX vocabulary journal

Month 3: Portfolio Beginning

- Complete second practice project (user research focus)
- Create a basic portfolio website
- Connect with 5 UX professionals for informational interviews
- Join a local UX meetup group

Month 4: Community Integration

- Volunteer UX help for nonprofit organization
- Complete third portfolio project (full design process)

- Attend in-person UX events
- Start sharing UX insights on LinkedIn

Month 5: Professional Preparation

- Polish portfolio with 3 strong projects
- Practice UX interview questions
- Develop 30-second elevator pitch
- Research target companies

Month 6: Active Job Search

- Apply for junior/entry UX positions
- Network actively at UX events
- Consider UX-adjacent roles (coordination, research assistant)
- Continue skill development based on interview feedback

Each step builds capability while providing validation and course correction opportunities.

The Resource Constraint Innovation Method

Use limitations as creative drivers:

> **Implementation Prompt:** "I want to achieve [goal] but have these constraints: [list limitations]. Help me turn each constraint into a creative advantage and generate solutions that work within these boundaries."

Example: The $100 Business Challenge

> **You:** "I want to start a business but have almost no money."
>
> **Claude:** "Perfect! Constraints breed creativity. With $100 total:

Service-Based Options (minimal upfront costs):

- Consulting: Domain name ($12) + basic website template (free) + business cards ($20) = $32 invested

- Digital products: Creation costs only time + $10 hosting = $10 invested
- Educational content: Free platform setup + $25 promotion budget = $25 invested

Creative Funding Approaches:

- Pre-sell services to fund operations
- Barter skills for needed services
- Use library resources for workspace and research
- Partner with others who have complementary resources

Growth Strategy:

- Start with one client, reinvest profits
- Document everything for content marketing
- Build referral incentives from day one"

Innovation Through Systematic Creative Techniques

Structured Innovation Questions

Use these systematic prompts to generate innovative thinking:

Possibility Expansion:

- "What would this look like if it were 10 times easier?"
- "How would a 5-year-old approach this problem?"
- "What would I do if failure were impossible?"
- "How can I make this more fun or engaging?"
- "What would be the opposite approach?"
- "How did nature solve similar challenges?"

Constraint Reframing:

- "How can this limitation become an advantage?"
- "What if I had to solve this with half the resources?"
- "How would I approach this if I only had one week?"
- "What if I could only use materials readily available?"

Perspective Shifting:

- "How would [different industry/culture] solve this?"
- "What would this look like from the user's perspective?"
- "How would I explain this to someone from 100 years ago?"
- "What would the critics of this idea focus on?"

Implementation Framework for Daily Creative Practice

Morning Creative Prompt (5 minutes):

> **Prompt Template:** "Give me a random creative challenge I can think about during my day: something that combines [two unrelated concepts] or asks 'what if [unexpected scenario]?'"

Afternoon Problem Reframe (5 minutes):

> **Prompt Template:** "Help me find a new solution to this recurring problem in my life: [describe issue]. Suggest 3 approaches I haven't tried."

Evening Creative Synthesis (5 minutes):

> **Prompt Template:** "I learned about [topic/experience] today. Help me connect this to [ongoing project/goal] in unexpected ways."

Creative Collaboration Best Practices

Maximize the effectiveness of your creative collaboration with Claude:

Effective Prompting for Creative Work

Be Specific About Creative Goals:

- Poor: "Give me creative ideas"
- Good: "I want to combine my expertise in [specific area] with my interest in [different area]"
- Better: "How can I create [specific type of content/solution] that appeals to [specific audience] while expressing [particular values/themes]?"

Request Multiple Approaches:

- "Generate 5 different approaches to [problem], each using a different creative strategy"
- "Show me conventional, unconventional, and completely experimental solutions to [challenge]"
- "What would [specific expert/creative person] do with this problem?"

Build Systematically:

- "Let's develop this idea step by step. First, help me understand the core appeal..."
- "Now that we have the basic concept, how can we make it more distinctive?"
- "What would make this idea more feasible to implement?"

Iteration and Development Process

First Round: Generate broad possibilities without constraints

Second Round: "Of these ideas, which 3 have the most potential? Why?"

Third Round: "Let's develop the most promising idea. What would success look like?"

Fourth Round: "What are the biggest obstacles to implementing this? How might we overcome them?"

Fifth Round: "What's the smallest version of this we could test quickly?"

Managing Creative Collaboration Boundaries

What Claude Does Well:

- Pattern recognition and connection-making
- Rapid generation of alternatives
- Research and information synthesis
- Systematic exploration of possibility spaces
- Consistent availability for iteration

What Requires Human Judgment:

- Aesthetic and artistic decisions
- Cultural sensitivity and appropriateness
- Feasibility assessment based on real-world constraints
- Emotional resonance and personal meaning
- Market timing and business validation

What to Verify Independently:

- Cultural references and accuracy
- Technical feasibility claims
- Market demand assumptions
- Legal or regulatory implications
- Resource requirement estimates

When Creative Collaboration Needs Redirection

Recognizing Unproductive AI Responses

Warning Signs:

- Generic suggestions that could apply to any problem
- Overuse of buzzwords without specific substance
- Recommendations that ignore stated constraints
- Ideas that sound impressive but lack practical implementation paths
- Responses that seem disconnected from your specific context

Recovery Strategies:

- **Simplify the Request:** "Let's focus on just one aspect of this problem: [specific element]"
- **Add Constraints:** "Given that I have [specific limitations], what's realistic?"
- **Request Examples:** "Show me one concrete example of this approach working"
- **Change Perspective:** "Approach this from a completely different angle"

Recovering from Creative Dead Ends

When Brainstorming Feels Stuck:

- **Switch Methods:** Move from analytical to emotional, or from individual to social focus
- **Change Scale:** Zoom out to bigger picture or zoom in to tiny details
- **Add Randomness:** "Introduce a completely random element to this problem"
- **Take Breaks:** Sometimes the most creative thing is stepping away

When Ideas Seem Unoriginal:

- **Dig Deeper:** "What makes this approach unique to my specific situation?"
- **Combine Concepts:** "How can I merge this idea with something from a completely different field?"
- **Personal Connection:** "How does this connect to my personal experiences or values?"

Knowing When to Abandon AI Assistance

Trust Your Human Intuition When:

- Ideas feel emotionally disconnected from your vision
- Suggestions ignore important cultural or social context
- Creative direction feels generic rather than personal
- You need to work through emotional or psychological aspects
- The problem requires deep domain expertise AI lacks

Return to Pure Human Creativity For:

- Deeply personal artistic expression
- Culturally sensitive creative decisions
- Problems requiring real-world experience
- Emotional or therapeutic creative work
- Breakthrough innovations requiring paradigm shifts

Real-World Creative Collaboration Examples

Case Study Approach to Creative Projects

Note The following represent approaches and methodologies rather than verified individual outcomes. Creative collaboration results vary significantly based on user skills, project complexity, and implementation quality.

Creative Challenge: Content Creator Struggling with Originality

The Systematic Approach:

1. **Problem Analysis:** Used Claude to identify content patterns and gaps in their niche
2. **Perspective Expansion:** Explored how other industries approach similar challenges
3. **Unique Angle Development:** Combined personal expertise with underexplored topics
4. **Content Planning:** Created a systematic approach to consistent, distinctive content

Key Success Factors:

- User brought deep domain knowledge and audience understanding
- Systematic iteration improved initial AI suggestions
- Clear creative vision guided collaboration decisions
- Human judgment filtered and refined AI-generated concepts

Creative Challenge: Small Business Differentiation

The Systematic Approach:

1. **Market Analysis:** Used Claude to research industry conventions and assumptions
2. **Value Proposition Exploration:** Systematically explored customer pain points

3. **Service Innovation:** Developed unique approaches to common business challenges
4. **Implementation Planning:** Created a feasible timeline with resource constraints

Key Success Factors:

- Business owner provided essential market and customer insights
- AI assistance helped identify overlooked opportunities
- Constraint-based thinking led to innovative solutions
- Real-world validation guided final decisions

Your Creative Collaboration Action Plan

Ready to systematically develop your creative collaboration skills? Here's your framework:

Week 1: Foundation Building

☐ Practice the systematic brainstorming method with three different problems

☐ Test the design thinking framework on one personal challenge

☐ Experiment with constraint-based creative thinking

☐ Start daily creative prompt routine

☐ Document what types of Claude responses work best for you

Week 2: Skill Development

☐ Complete one creative project using the full collaboration framework

☐ Practice iteration and refinement with AI feedback

- ☐ Experiment with different prompting approaches
- ☐ Learn to recognize and redirect unproductive AI responses
- ☐ Develop personal creative collaboration guidelines

Week 3: Advanced Application

- ☐ Apply strategic planning framework to one significant goal
- ☐ Use innovation questions to approach recurring challenges
- ☐ Practice perspective-shifting techniques
- ☐ Test the stepping stone strategy on one area of desired change
- ☐ Refine your personal creative collaboration process

Week 4: Integration and Evaluation

- ☐ Complete one substantial creative project from start to finish
- ☐ Evaluate which collaboration techniques work best for your style
- ☐ Share creative work for feedback and learning
- ☐ Plan ongoing creative development goals
- ☐ Establish sustainable creative practice using AI assistance

The Innovation Mindset: Systematic Creative Development

Developing an **innovation** mindset isn't about becoming a different person. It's about recognizing the creative problem-solver you already are and developing systematic approaches to enhance that natural capability.

Questions That Spark Systematic Innovation

Use these as regular creative prompts:

Possibility Questions:

- "What would this look like if it were easy?"
- "How would a 5-year-old solve this?"
- "What would I do if I knew I couldn't fail?"
- "How can I make this more fun?"
- "What would the opposite approach be?"
- "How did nature solve similar problems?"

Implementation Strategy: Choose one question weekly and apply it to multiple challenges. Track which questions consistently generate useful insights for your thinking style.

Daily Creative Practice Framework

Just five minutes daily builds creative muscle systematically:

Morning (2 minutes): Ask Claude for a random creative prompt connecting two unrelated concepts

Afternoon (2 minutes): Find a new solution to a routine problem using constraint-based thinking

Evening (1 minute): Identify one unexpected connection from your day's experiences

Weekly Review: Notice patterns in what sparks your best creative thinking

Common Creative Collaboration Pitfalls (and Systematic Solutions)

The Complexity Trap

Problem: Making creative challenges harder than necessary

Solution Framework: "Help me simplify this creative challenge to its essential elements. What's the core problem if we remove all unnecessary complexity?"

The Originality Obsession

Problem: Dismissing ideas as "not original enough"

Solution Framework: Remember—execution and personal perspective create originality. Ask: "How does my unique context/ experience change this familiar approach?"

The Perfect Timing Myth

Problem: Waiting for the "right time" to start creative projects

Solution Framework: "What's the smallest creative step I could take right now with exactly the resources I have today?"

The Solo Genius Fantasy

Problem: Thinking creativity must happen in isolation

Solution Framework: Embrace systematic collaboration—with Claude for ideation and with humans for validation and implementation guidance.

The Bottom Line

Creativity isn't a mystical gift bestowed upon the chosen few. It's a capability you can develop systematically by showing up, applying structured approaches, and being willing to iterate. With Claude as your creative collaboration partner, you have access to pattern recognition and information synthesis that can enhance your natural creative problem-solving abilities.

The key insights for effective creative collaboration:

Understand AI's Role: Claude provides pattern-based suggestions and systematic exploration, not genuine creative insight. The creative vision and judgment remain yours.

Use Systematic Approaches: Structured frameworks like design thinking and strategic planning make creativity more reliable and productive.

Iterate Purposefully: Each round of collaboration should build understanding and move toward actionable solutions.

Validate Through Reality: Test ideas with real people in real situations to distinguish between clever concepts and practical solutions.

Maintain Human Ownership: You provide the vision, values, and judgment that transform AI assistance into meaningful creative work.

Remember: Every innovation started as someone's systematic exploration of "What if?" Every creative breakthrough came from someone willing to combine existing elements in new ways. You already solve problems creatively every day. Claude is here to amplify those abilities and help you work more systematically.

The world needs what you uniquely can create. Not because you're a creative genius (though you might be), but because you're you. Your perspective, your experience, and your systematic approach to combining influences—that's where innovation lives.

So stop waiting for inspiration. Stop comparing yourself to others. Stop thinking you need permission to create. Use the systematic frameworks in this chapter, collaborate purposefully with Claude, and start building something meaningful. Even if your first attempt isn't perfect, you'll have learned something valuable for the next iteration.

Ready to create systematically? Claude's ready when you are. Let's innovate!

CHAPTER 9

Data Analysis and Visualization

In This Chapter

- Discovering you're already a data analyst (surprise!)
- Making sense of numbers without a math degree
- Turning boring spreadsheets into "aha!" moments
- Understanding statistics without breaking into a cold sweat
- Creating visualizations that actually visualize
- Building business intelligence from everyday information

Last month, you probably analyzed more data than a 1990s Fortune 500 company. Your fitness tracker counted steps, your banking app categorized spending, your streaming service tracked viewing habits, and your brain processed it all without a single statistics course. The only difference between you and a "data analyst" is they get paid to do what you do naturally. With Claude as your analytical partner, you're about to discover that **statistics** tell stories, **pattern recognition** is your superpower, and creating **visualizations** beats staring at endless numbers every time.

A Framework for AI-Assisted Data Analysis

Before diving into techniques, let's establish a systematic approach that connects data analysis principles to Claude collaboration:

R. Dickey, *Mastering Claude AI*, https://doi.org/10.1007/979-8-8688-2001-4_9

Step 1: Define Your Analysis Goal

- What question are you trying to answer?
- What decision will this data help you make?
- What level of accuracy do you need?

Step 2: Assess Your Data Quality

- Check for missing values, outliers, and inconsistencies
- Verify data sources and collection methods
- Understand potential biases or limitations

Step 3: Choose Appropriate Analysis Methods

- Match the analysis type to your question (descriptive, comparative, trend analysis)
- Consider sample size requirements for reliability
- Plan for statistical validation when needed

Step 4: Interpret Results with Appropriate Caution

- Distinguish between correlation and causation
- Acknowledge limitations and confidence levels
- Validate insights through multiple approaches when possible

This framework connects to every technique we'll explore, ensuring your AI-assisted analysis remains methodologically sound.

The Data Analysis Myths (Let's Shred Them)

Time to destroy the myths that keep people from embracing their inner data analyst:

1. **Myth #1: "You need to be good at math"**
   Reality: You need to be curious about what numbers mean, not how to calculate them by hand.

2. **Myth #2: "Data analysis requires expensive software"**
   Reality: Claude can help interpret basic data patterns and guide analysis approaches, though it has limitations in performing complex calculations or processing large datasets. For sophisticated statistical work, you'll need dedicated tools, but many insights come from simple analysis.

3. **Myth #3: "Statistics are just lies in disguise"**
   Reality: Statistics are stories told by numbers. Like any story, they can be told well or poorly.

4. **Myth #4: "Only big companies need data analysis"**
   Reality: Anyone making decisions benefits from data—including you deciding which coffee shop has the best value.

What Is Data Analysis Really? (You're Already Doing It)

Data analysis is simply looking at information to understand what it means and make better decisions. Every time you:

- Compare product reviews before buying
- Track your exercise progress
- Budget your monthly expenses
- Notice your kid gets cranky at 3 PM every day
- Realize traffic is always worse on Thursdays

You're analyzing data! Claude can help you do it more systematically and find patterns you might miss, while you maintain responsibility for verifying insights and making decisions.

Your Daily Data (It's Everywhere)

Before we dive into techniques, let's recognize the data sources in your life while being mindful of privacy considerations:

Personal Data You Already Have

Note When sharing data with Claude, anonymize personal information and avoid including sensitive details.

- Bank statements (spending patterns)
- Fitness tracker info (health trends)
- Social media insights (what content works)
- Email patterns (when you're most productive)
- Sleep data (what affects your rest)
- Mood patterns (what triggers good/bad days)

Work Data Hiding in Plain Sight

Caution Ensure you have permission before sharing proprietary business information and follow your organization's data policies.

- Sales numbers (what's really selling)
- Customer feedback (common complaints)
- Time tracking (where hours really go)

- Email response rates (what subjects work)
- Meeting effectiveness (which ones matter)
- Project timelines (accurate estimates)

Pattern Recognition: Your Brain's Superpower

Pattern recognition—seeing connections and trends in information—is something humans excel at naturally. Claude can help amplify this ability by processing structured data and suggesting analytical approaches, but the insights still require your interpretation and validation.

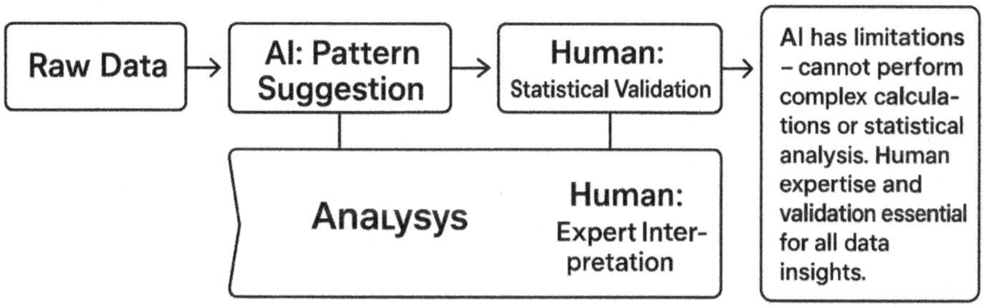

Figure 9-1. *From raw data to informed decisions: Pattern recognition illuminates insights hidden in the numbers, transforming information overload into actionable intelligence*

The Coffee Shop Example

You: "Claude, I tracked my coffee purchases for a month. Help me understand the patterns."

Implementation Framework:

> Prompt Template: "I have [time period] of data about [specific behavior].
>
> The data includes [list key variables]. Help me identify potential
>
> patterns and suggest what additional factors I should consider."

Claude's Systematic Analysis Approach:

- Frequency patterns (daily vs. weekend habits)
- Cost patterns (price sensitivity indicators)
- Time correlations (when you spend most)
- External factors to consider (weather, stress, schedule)

Expected Outcome: Pattern identification with appropriate caveats about sample size and potential confounding variables.

Making Sense of Numbers (No PhD Required)

Let's explore five fundamental types of analysis that Claude can help you understand:

1. Averages: Finding Your Baseline

You: "What's my typical grocery spending?"

Claude's Guidance: "Let's calculate your baseline spending and examine the distribution. I can help you identify outliers and understand what 'typical' means in your context."

Implementation Note: Claude can guide calculation approaches, but verify mathematical results independently for important decisions.

2. Trends: Seeing Direction Over Time

You: "Is my morning run time improving?"

Claude: "Based on your data, you've reduced your average time from 20 minutes in January to 18 minutes now. This shows improvement, though we should note the data spans only [X] observations."

Critical Qualification: Trend identification requires sufficient data points and should account for seasonal or other cyclical factors.

3. Comparisons: Understanding Relationships

You: "Do I work out more on weekdays or weekends?"
Claude: "Your data shows weekday average of 4 workouts vs. weekend average of 1.5. However, this comparison assumes similar opportunity on both weekday and weekend days."

4. Frequency: How Often Things Happen

You: "How often do clients actually call vs. email?"

Claude: "87% email, 13% calls. But call clients spend 3x more. This suggests different communication preferences might correlate with different client values."

5. Correlation: Things That Happen Together

You: "Is there a connection between my sleep and productivity?"

Claude: "Your data suggests a potential relationship: 7+ hours sleep correlates with 85% task completion vs. 45% with less than 6 hours. However, establishing true correlation requires statistical testing and sufficient sample size. Simple pattern observation should be verified through additional analysis methods."

Critical Statistical Note: Correlation analysis requires proper methodology, adequate sample sizes, and consideration of confounding variables. Pattern recognition alone is not a statistically reliable correlation.

Making Statistics Personal

Prompt Framework for Personal Analysis:

> "I tracked [specific behavior/metric] for [time period]. Here's my data: [dataset].
>
> Help me find meaningful patterns while noting any limitations in sample size
>
> or methodology. What additional data might strengthen these insights?"
>
> **You:** "I tracked my mood (1-10) and activities for a month. Help me find what makes me happier."

Claude's Systematic Analysis:

- **Exercise Days**: Average mood 7.8
- **No Exercise**: Average mood 5.9
- **Social Evenings**: +1.5 mood boost next day
- **Nature Time**: Consistent 8+ mood scores
- **Deadline Days**: -2 mood drop (plan recovery)

> **Statistical Qualification:** "These patterns are based on [X] data points over [Y] period. Consider tracking for additional months to verify patterns and account for seasonal or other variables."
>
> **The Insight:** Numbers tell your story, but always with appropriate context about reliability and limitations.

Visualization: Making Data Come Alive

Visualization is just a fancy word for "showing information in pictures." Our brains process visuals 60,000 times faster than text. Claude can help you conceptualize and plan visualizations, though you'll need other tools for actual chart creation.

The Right Chart for the Right Story

Pie Charts: Parts of a Whole

You: "Show me where my time goes in a typical day"

Claude's Visualization Guidance: "A pie chart would effectively show time allocation:

- **Sleep**: 30% (7.2 hours)
- **Work**: 35% (8.4 hours)
- **Commute**: 8% (2 hours)
- **Meals**: 8% (2 hours)
- **Family**: 12% (2.9 hours)
- **Personal**: 7% (1.7 hours)"

Line Graphs: Changes Over Time

You: "How has my savings changed this year?"

Claude: "A line graph would best illustrate your savings trajectory: starting at $1,000 in January, dipping to $800 in March (car repair), then climbing steadily to $2,400 by December. The upward trend demonstrates financial progress despite setbacks."

Bar Charts: Comparing Categories

You: "Which blog posts get the most engagement?"

Claude's comparative analysis:

- **How-to Posts**: 450 average views
- **Personal Stories**: 380 views
- **Industry News**: 150 views
- **Product Reviews**: 600 views (winner!)

DIY Visualization with Claude

You: "I want to visualize my project timelines but can't use complex software."

Claude: "Let's create a simple text-based Gantt chart representation:

Project Timeline (Weeks 1–8):

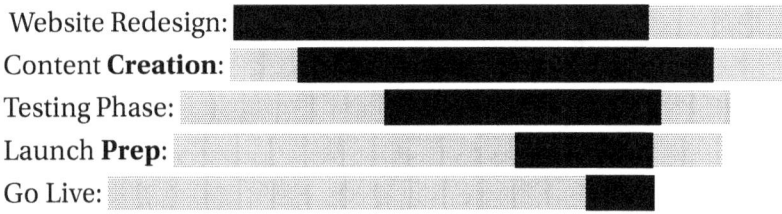

Simple, clear, and you can create it in 30 seconds! For actual project management, you'll want dedicated software, but this helps visualize the concept."

Business Intelligence for Normal Humans

Business intelligence sounds corporate, but it's really just using data to make smarter decisions. You don't need a Fortune 500 company to benefit.

Small Business Intelligence

The Freelancer's Dashboard:

Prompt Framework:

"I'm a [type of professional]. What key metrics should I track for business insight?

Help me understand which measurements would provide actionable intelligence for someone in my situation."

You: "I'm a freelance designer. What data should I track?"

Claude's Strategic Suggestions:

- Client acquisition cost (marketing spend ÷ new clients)
- Average project value (spot pricing opportunities)
- Time per project type (find your profitable niches)
- Client lifetime value (who to nurture)
- Busy/slow seasons (plan accordingly)
- Payment delays by client (cash flow management)

Personal Life Intelligence

Family Budget Analysis:

You: "Help me understand our spending without complex budgeting apps."

Claude's Systematic Approach: "Track just five categories for one month:

1. Fixed costs (rent, insurance)
2. Food (groceries + dining)
3. Transportation
4. Fun money
5. Everything else

After One Month, We Can Analyze:

- Surprise spending areas
- Easy savings opportunities
- Realistic budget targets
- Progress tracking methods"

Real-World Data Analysis Projects:

Let's walk through actual analyses you can do today, with proper methodological frameworks:

Project 1: The Energy Audit

Goal: Reduce utility bills

Implementation Framework:

1. **Data Collection:** Gather 12 months of utility bills
2. **Quality Check:** Verify all bills are complete and comparable
3. **Analysis Approach:** Identify patterns while noting limitations
4. **Action Planning:** Develop testable hypotheses

   You: "Here are my last 12 electric bills: [amounts]"

Claude's Systematic Analysis:

- Seasonal patterns (summer 60% higher than baseline)
- Baseline usage identification (minimum monthly cost)
- Unusual spikes (December requires investigation)
- Year-over-year comparison methodology
- Estimated savings potential with confidence intervals

Action Plan: Target summer cooling efficiency; investigate the December anomaly; baseline analysis suggests $30/month potential savings (±$10 margin).

Project 2: The Productivity Detective

Goal: Find your peak performance times

Methodology Framework:

1. **Data Collection Protocol:** Rate energy (1–10) every hour for two weeks minimum

2. **Control Variables:** Note sleep, meals, activities, and external factors
3. **Analysis Limitations:** Acknowledge that self-reporting can be subjective
4. **Validation Approach:** Test insights through schedule adjustments

Claude Helps You Discover:

- Peak performance time patterns (with sample size considerations)
- Post-meal energy correlations (noting individual variation)
- Day-of-week patterns (accounting for schedule differences)
- Task-type energy requirements
- Optimal break timing based on your data

Result: Schedule adjustments matched to your specific patterns, with built-in testing to validate insights.

Advanced Analysis Made Simple:

Predictive Patterns (No Crystal Ball Required)

You: "Based on my sales data, what should I expect next month?"

Claude's Response with Appropriate Limitations: "Based on your data patterns, I can identify trends, but reliable forecasting requires sophisticated modeling beyond simple pattern recognition. Here's what your data suggests:

- **Monthly Growth Trend**: 12% average (based on [X] months)
- **Seasonal Factor**: +20% in December (if pattern holds)
- **Current Trajectory**: Consistent upward trend
- **Suggested Range**: $4,200-4,800 (based on pattern continuation)
- **Important Limitations**: This is pattern extrapolation, not statistical forecasting. External factors, market changes, or seasonal variations could significantly impact actual results."

Critical Qualification: Simple pattern extrapolation is not reliable forecasting. True predictive analysis requires sophisticated modeling, validation techniques, and consideration of multiple variables beyond identifying trends.

Comparative Analysis (Apples to Apples)

You: "How does my blog performance compare to last year?"

Claude's Systematic Comparison Framework:

Metric	Last Year	This Year	Change	Context
Posts	24	52	+117%	Doubled output
Avg Views	125	310	+148%	Strong growth
Comments	2.1	5.7	+171%	Better engagement
Time/Post	4 hrs	2.5 hrs	-38%	More efficient

Insight: You're writing more, faster, and better. The data demonstrates clear improvement across multiple metrics, suggesting both increased efficiency and audience growth.

Segmentation (Finding Hidden Groups)

You: "My customer feedback seems all over the place."

Claude's Pattern Analysis: "I can help identify potential customer segments based on feedback themes, though true segmentation analysis requires statistical clustering methods. Based on your feedback data, I see three possible groups:

1. **Price-Sensitive (30%)**: Frequently mention deals, sales, and discounts
2. **Quality-Focused (45%)**: Emphasize craftsmanship, durability, and materials

3. **Convenience-Seekers (25%)**: Prioritize speed, ease, and accessibility

Important Note: This is preliminary pattern recognition. Validate these segments through additional data collection and statistical analysis before making significant business decisions."

Common Data Analysis Pitfalls:

The Correlation/Causation Trap

You: "Ice cream sales and crime both go up in summer. Ice cream causes crime?"

Claude: "This is a classic correlation/causation error! Both increase due to a third factor—hot weather brings more people outside, creating both higher ice cream sales AND more opportunities for crime. Always ask: 'What other factors could explain this relationship?' True causation requires controlled studies and elimination of confounding variables."

The Cherry-Picking Problem

You: "My best month ever was $10K in sales!"

Claude: "Congratulations! However, for meaningful analysis, we need to look at the full picture. What's your average monthly performance? How does this compare to your typical range? One exceptional month doesn't establish a trend—we need to understand whether this was an outlier or part of a sustainable pattern."

The Too-Much-Data Paralysis

You: "I have 50 metrics to track!"

Claude: "Start with 3-5 metrics that directly impact your primary goals. You can always add more later. Quality analysis of key metrics beats superficial tracking of everything. Focus on actionable insights rather than comprehensive measurement."

The Perfect-Data Procrastination

You: "I'll start analyzing when I have better data..."

Claude: "Start with what you have now. Imperfect data today provides learning opportunities and insights that beat perfect data never collected. You'll improve your data quality and analysis skills through practice."

Data Quality and Validation Framework:

Essential Data Quality Checks:

Before any analysis, implement this systematic quality assessment:

1. **Completeness Check**
    - Are there missing data points?
    - Do gaps follow patterns (weekends, holidays)?
    - How will missing data affect conclusions?
2. **Accuracy Verification**
    - Are values within reasonable ranges?
    - Do extreme values represent errors or real events?
    - Can outliers be verified through additional sources?

3. **Consistency Review**
    - Are measurements taken using the same methods?
    - Have definitions or criteria changed over time?
    - Are data sources comparable?
4. **Bias Assessment**
    - How was data collected?
    - Who was included/excluded?
    - What circumstances might skew results?

Prompt Framework for Quality Assessment:

> "Help me assess the quality of this dataset: [describe data].
> What potential issues should I check for? What limitations
> Should I acknowledge when interpreting results?"

Statistical Rigor Guidelines:

When to Seek Statistical Validation

Consider formal statistical analysis when

- Making important business or personal decisions
- Sample sizes are small (generally under 30 observations)
- Patterns seem inconsistent or contradictory
- External validation is required
- Stakes are high for incorrect conclusions

Sample Size Considerations

Claude's Guidance: "For reliable insights, consider these general guidelines:

- **Trend Analysis**: Minimum 12–20 data points
- **Comparison Studies**: At least 30 observations per group

- **Correlation Analysis**: 50+ paired observations recommended
- **Complex Pattern Recognition**: 100+ observations preferred

Remember: Larger samples generally provide more reliable insights, but quality matters more than quantity."

Confidence and Uncertainty

Always acknowledge limitations in your analysis:

- "Based on [X] observations over [Y] time period..."
- "This pattern appears consistent, though additional data would strengthen conclusions..."
- "These insights should be validated through..."
- "Consider these potential confounding factors..."

Failure Recovery Framework

When Data Analysis Gets Stuck

Pattern Recognition Failure:

- Switch analytical approaches (detailed to overview, or vice versa)
- Add different time periods or groupings
- Consider external factors not initially included
- Take breaks—fresh perspectives often reveal patterns

When Insights Seem Unclear:

- Return to your original question—are you analyzing the right data?
- Simplify your analysis—complex isn't always better
- Seek patterns at different scales (daily vs. monthly vs. yearly)
- Cross-reference with qualitative observations

When to Abandon AI Assistance:

- Analysis requires specialized statistical expertise
- Data involves sensitive personal or business information
- Results have significant legal or safety implications
- Domain expertise trumps pattern recognition

Validation Through Multiple Approaches

Always Verify Important Insights:

1. **Replicate findings** using different time periods
2. **Cross-reference** with external benchmarks
3. **Test predictions** through small-scale experiments
4. **Seek feedback** from domain experts
5. **Monitor results** of decisions based on analysis

Your Data Analysis Toolkit:

Essential Tools You Already Have

Built-In Phone Tools

- Notes app (quick data capture)
- Calculator (basic math verification)
- Camera (photograph receipts, whiteboards)
- Voice memos (capture insights)
- Calendar (time tracking)

Free Basics

- Google Sheets (simple spreadsheets and charts)
- Claude (analysis guidance and interpretation)

- Email (data collection via forms)
- Your brain (pattern recognition and validation)

The Claude Advantage

- No formula memorization required
- Plain English analysis guidance
- Pattern recognition assistance
- Methodology suggestions
- Next-step recommendations

Important Limitation: Claude cannot execute formulas, perform complex calculations, or process large datasets directly. Use it for guidance, interpretation, and methodology—verify calculations independently.

Making It Stick: Your 30-Day Data Challenge:

Week 1: Foundation

- ☐ Pick one simple thing to track (mood, spending, time)
- ☐ Establish data quality protocols
- ☐ Collect data daily without analysis
- ☐ Notice how awareness alone creates change

Week 2: Basic Analysis

- ☐ Share week 1 data with Claude for pattern identification
- ☐ Find three potential patterns
- ☐ Calculate one meaningful average
- ☐ Create one simple visualization concept

Week 3: Deeper Insights

- ☐ Add a second data type
- ☐ Look for correlations with appropriate caution
- ☐ Make one testable prediction
- ☐ Identify potential confounding factors

Week 4: Action and Validation

- ☐ Make one decision based on data insights
- ☐ Track the results of your decision
- ☐ Share learnings with someone else
- ☐ Plan your next analysis with improved methodology

Quick Wins: Analyses You Can Do in 10 Minutes

The Email Efficiency Check:

Examine your sent folder timestamps for one week. You'll likely discover your peak email efficiency times—perhaps sharp and quick from 9 to 11 AM, but sluggish after 2 PM. Schedule important emails during high-energy periods.

Statistical note: One week provides initial insights; track for a month to verify patterns.

The Time Truth Bomb:

Track actual time use for one day in 30-minute increments, then compare to your perceived time allocation. Most people discover significant gaps between perception and reality.

Methodology note: Single-day tracking provides awareness; weekly tracking enables pattern identification.

The Happiness Audit:

> Rate your mood (1-10) after different activities for three days. Quickly identify energy-giving vs. energy-draining activities.
>
> Validation approach: Test insights by intentionally scheduling more high-energy activities and measuring results.

The Bottom Line

Data analysis isn't about becoming a statistician or learning complex software. It's about being systematically curious about patterns in your life and using methodologically sound approaches to understand them better. With Claude as your analysis partner, you can transform random information into actionable insights while maintaining appropriate skepticism about results.

You're already swimming in useful data—from daily routines to business metrics to personal goals. The difference between drowning in numbers and surfing insights is having the right methodology, a good thinking partner, and healthy skepticism about conclusions.

Remember: Every successful person and business uses data to make decisions. They're not necessarily smarter or more mathematical than you—they follow systematic approaches, acknowledge limitations, and validate insights before acting. With Claude's guidance and these methodological frameworks, you can do the same.

The goal isn't to become a data scientist. It's to become systematically data-informed. To notice patterns that help you save money, save time, make better decisions, and understand what actually works in your life and work—while always maintaining appropriate caution about the reliability and limitations of your insights.

Start small. Pick one question you want answered. Collect simple data using quality protocols. Share it with Claude for pattern guidance. Find insights with appropriate skepticism. Test conclusions through action. Refine your approach. Before you know it, you'll be that person who "just knows" when things will work out—except you'll know because you have methodologically sound data to back it up.

Ready to discover what your data has been trying to tell you? Open a spreadsheet, start a list, or just begin noticing patterns systematically. Claude's ready to help you translate numbers into knowledge and knowledge into action—with all the appropriate caveats and validation steps that make insights reliable. Let's analyze!

PART III

Advanced Techniques

CHAPTER 10

Advanced Prompting Strategies

In This Chapter

- Recursive chain of thought that builds on itself
- Meta-prompting loops that optimize themselves
- Constraint engineering for breakthrough thinking
- Understanding technique limitations and failure modes
- Building decision frameworks for advanced prompt selection

You've mastered the basics. You can write a decent prompt, get Claude to role-play, and even show its work. But now you're ready for the advanced techniques that can enhance Claude's problem-solving capabilities while understanding their appropriate limitations. These aren't your Chapter 4, "The Art of Prompting: Getting Better Responses," prompting tips—these are sophisticated methods that experienced users employ when tackling complex challenges, with realistic expectations about what they can and cannot achieve.

Important Note Advanced prompting techniques improve AI output quality but cannot overcome fundamental limitations in AI reasoning, knowledge, or problem-solving capabilities. These methods help you work more systematically with AI's pattern-recognition abilities, not replace human expertise or critical thinking.

R. Dickey, *Mastering Claude AI*, https://doi.org/10.1007/979-8-8688-2001-4_10

A Framework for Advanced Prompting Strategy

Before diving into specific techniques, here's a systematic approach for choosing and applying advanced prompting methods:

Step 1: Problem Assessment

- **Complexity Level**: Simple question, multi-step analysis, or open-ended exploration?
- **Required Accuracy**: Low-stakes brainstorming or high-stakes decision support?
- **Time Available**: Quick iteration or thorough exploration?
- **Your Expertise**: Domain familiarity or learning new territory?

Step 2: Technique Selection

- **Single-Pass Problems**: Use basic prompting
- **Layered Analysis Needed**: Consider recursive chain of thought
- **Process Optimization**: Try meta-prompting loops
- **Creative Block-Breaking**: Apply constraint engineering

Step 3: Implementation with Reality Checks

- Start with a simplified version
- Monitor for error amplification
- Validate outputs through independent means
- Know when to abandon complex approaches

Step 4: Quality Control

- Recognize ineffective outputs early
- Apply fallback strategies when needed
- Use human judgment for final evaluation

Recursive Chain of Thought: The Thinking That Thinks About Thinking

Remember the basic chain of thought from Chapter 4, "The Art of Prompting: Getting Better Responses"? That's like arithmetic. This is calculus. Recursive chain of thought doesn't just show reasoning—it builds layers of analysis that reference and refine each other, though it's important to understand this can also amplify errors if built on incorrect foundations.

Understanding the Risks and Benefits

What Recursive Prompting Does Well:

- Creates structured analytical progression
- Forces examination of assumptions
- Provides transparency in reasoning steps
- Helps break down complex problems systematically

Critical Limitations to Understand:

- Can amplify initial errors through each iteration
- May create illusion of depth without genuine insight
- Generates pattern-based responses, not true logical analysis
- Requires human validation at each critical step

The Recursive Loop Technique

Note This technique works most effectively when you need a systematic breakdown of complex problems, though results require verification for critical decisions.

Instead of linear thinking, create loops that spiral deeper:

"Analyze this business problem using recursive thinking:

Level 1: Identify the core issue

Level 2: Question your Level 1 assumptions—what might you be missing?

Level 3: Synthesize Levels 1 and 2 into a refined understanding

Level 4: Find the pattern between your initial and refined understanding

Level 5: Apply that pattern to predict what you haven't considered yet

Show your thinking at each level, explicitly referencing previous levels.

Important After each level, pause to verify accuracy before proceeding. If any level contains errors, they will compound in subsequent levels."

The Self-Questioning Cascade

This technique makes Claude interrogate its own responses with built-in quality checks:

"Solve this problem: [YOUR PROBLEM]

After Your Initial Solution:

- Ask yourself three critical questions about your approach
- Answer those questions

- Based on those answers, ask three deeper questions
- Answer those
- Now revise your original solution
- Explain what changed and why

Validation Check: *If at any point the reasoning seems circular or the questions become irrelevant, simplify the approach and focus on the core issue."*

Real-World Recursive Application

The Strategic Decision Spiral

Here's how to apply this systematically:

"Should we expand our product line or improve existing products? Use recursive analysis:

> **Round 1**: Initial recommendation with reasoning
>
> **Round 2**: Challenge three assumptions from Round 1
>
> **Round 3**: Integrate challenges into refined recommendation
>
> **Round 4**: Project second-order effects of Round 3 recommendation
>
> **Round 5**: Final recommendation accounting for all rounds

Quality Control: *After Round 3, validate key assumptions through external research. If major contradictions emerge, restart with simpler approach."*

Implementation Success Story

Sarah Used This for a Life-Changing Decision, but with Realistic Expectations:

"Should I move across the country for this job opportunity?

> **Level 1**: List obvious pros and cons
>
> **Level 2**: Question what's behind each pro and con—what am I really valuing?
>
> **Level 3**: Based on those values, what matters most?
>
> **Level 4**: What pattern do I see in how I make major decisions?
>
> **Level 5**: How does recognizing this pattern change my choice?

Sarah's Result: *The recursive analysis helped her recognize her decision-making patterns, though she validated the insights with trusted advisors before making her choice."*

Meta-Prompting Loops: Systematic Process Improvement

This isn't just asking Claude to help you prompt better. This is creating feedback loops where you systematically refine the conversation process, understanding that AI cannot truly optimize but can help you identify improvement patterns.

Understanding Meta-Prompting Realities

What Meta-Prompting Actually Does:

- Helps identify effective prompt patterns through experimentation
- Provides a structured approach to iteration
- Offers feedback on prompt effectiveness patterns
- Creates systematic improvement methodology

Important Limitations:

- AI cannot truly "optimize"—it responds to prompts about improvement
- Humans must evaluate and guide actual improvement
- Quality assessment requires clear, objective criteria
- Success depends on the user's ability to recognize effective outputs

The Prompt Evolution Engine

"I need to solve [COMPLEX PROBLEM].

> **Step 1**: What are the 5 most important questions I should be asking about this?
>
> **Step 2**: Answer question #1

Step 3: Based on that answer, how should I refine questions 2–5 to be more targeted?

Step 4: Answer the refined question #2

Step 5: Continue this pattern, letting each answer refine the remaining questions

Step 6: What crucial question did we never ask but should have?

Quality Check: *After Step 3, evaluate whether the refined questions are actually more useful than the original ones. If not, adjust the approach."*

The Systematic Improvement Loop

"Let's systematically improve how we're approaching [TASK]:

Attempt 1: [Initial approach]

After your response, evaluate:

- What specific elements worked well?
- What could be improved and why?
- How should I adjust my next prompt for better results?

[Apply feedback in next prompt]

Attempt 2: [Refined approach based on feedback]

Continue until the output meets defined quality criteria. Track what improvements made the biggest difference.

Quality Assessment Framework: Rate outputs on specific criteria (e.g., clarity 1–10, completeness 1–10, actionability 1–10) rather than subjective judgments. Define what constitutes success before beginning the process."

Meta-Learning in Practice

Marcus revolutionized his writing process using meta-prompting with realistic expectations:

"Claude, help me write better marketing emails. After each draft, provide:

1. What specific elements made it effective or ineffective?
2. What one change would most improve it?
3. What pattern do you notice in my writing approach?

Let's do 5 rounds and track the improvement systematically.

Marcus's Implementation: He defined specific success metrics (open rates, click-through rates) and validated Claude's suggestions against his actual email performance data. By round 5, his emails weren't just better—he understood WHY they were better and could apply those principles independently."

Constraint Engineering: Creative Stimulation Through Boundaries

Advanced constraints don't automatically produce breakthroughs—they stimulate novel AI responses by creating productive friction, though breakthrough insights require human creativity and domain expertise to evaluate and implement.

Understanding Constraint-Based Creativity

What Constraints Do Effectively:

- Force exploration of unconventional solution spaces
- Prevent AI from defaulting to generic responses
- Create structured creative pressure
- Generate novel combinations from training patterns

Critical Reality Checks:

- Novel AI responses aren't automatically breakthrough insights
- Constraints may produce interesting but impractical suggestions
- Human expertise required to evaluate feasibility and value
- Implementation success depends on real-world domain knowledge

The Impossible Constraint Challenge

"Solve [PROBLEM] with these 'impossible' constraints:

- Zero budget
- No new technology
- Must work for opposing user groups
- Implement by tomorrow
- Creates no new problems

First Reaction: 'That's impossible' But then... find the creative solution hiding in these constraints. What assumption must you shatter to succeed?

Reality Check: *Evaluate each suggested solution for actual feasibility. Many 'impossible constraint' solutions sound clever but require significant modification for real-world implementation."*

The Cascading Constraint System

"Explain [COMPLEX TOPIC]:

Version 1: Full explanation (baseline)

Version 2: Half the words, same clarity

Version 3: Half again, maintain essence

Version 4: Reduce to a single metaphor

Version 5: Express as a question that contains its own answer

What survived all constraints? That's your core insight.

Implementation Note: This technique is most valuable for identifying essential elements, not for creating final explanations. Use insights from this exercise to improve your full explanation."

Real Constraint Applications

The Business Challenge Reframe

Jennifer's consultancy was struggling. Her constraint challenge:

"Save my business with:

- No marketing budget
- No sales team
- No new services
- Must be profitable in 30 days
- Can't work more hours"

The Process: *The "impossible" constraints forced exploration of overlooked resources (existing clients as referral sources). However, Jennifer had to validate feasibility, modify the profit-sharing approach for legal compliance, and create systematic implementation rather than just the initial idea.*

The Teaching Innovation Framework

A professor used cascading constraints to improve course design:

"Explain quantum mechanics:

Version 1: Traditional lecture (2000 words)

Version 2: Using only everyday objects (1000 words)

Version 3: As a children's story (500 words)

Version 4: In one drawing with labels (visual)

Version 5: As a single question: 'Why can't you know everything about anything?'"

Implementation Result: *The constraints revealed core concepts, but the professor spent additional time developing proper scientific accuracy and age-appropriate complexity for the actual course.*

The Master Technique: Systematic Advanced Integration

Here's where the techniques combine—combining all three approaches with realistic expectations and quality controls:

"Help me solve [MAJOR CHALLENGE]:

PART 1: Recursive Understanding

> **Level 1**: What's the obvious problem?
>
> **Level 2**: What's the problem behind the problem?
>
> **Level 3**: What's the problem we're not seeing?

Quality Check: *Validate each level before proceeding. If reasoning becomes circular, focus on the most concrete level.*

PART 2: Meta-Prompt Evolution

> Based on Part 1, what are the 3 most important questions we should be asking? Answer one, then revise the others based on what we learn.

Reality Check: *Ensure refined questions are actually more useful than original ones.*

PART 3: Constraint Breakthrough Now solve it with:

- Using only resources we already have
- In a way that our competition can't copy
- That gets easier as it scales
- Creates value for everyone involved

Implementation Filter: *Evaluate each constraint-based solution for real-world feasibility.*

PART 4: Systematic Validation Question your solution's assumptions. Test through small-scale experiments where possible. Seek feedback from domain experts. Refine based on practical constraints

What emerged that we couldn't see at the start? What requires further development before implementation?"

Your Advanced Mastery Checklist

Ready to level up with realistic expectations? Here's your systematic approach:

Week 1: Master Recursive Thinking

- ☐ Apply recursive chain of thought to one personal decision
- ☐ Use the self-questioning cascade on a work problem
- ☐ Document insights that only emerged in later levels
- ☐ Validate key insights through independent research
- ☐ Share the technique with someone who's stuck

Week 2: Optimize with Meta-Prompting

- ☐ Create a prompt evolution engine for your biggest challenge
- ☐ Define specific quality criteria before beginning
- ☐ Run an improvement loop with measurable outcomes
- ☐ Track which refinements made the biggest difference
- ☐ Build a personal library of effective prompt patterns

Week 3: Breakthrough with Constraints

- ☐ Apply "impossible" constraints to a stalled project
- ☐ Use cascading constraints to identify core elements
- ☐ Evaluate constraint-based solutions for practical feasibility
- ☐ Implement one insight from constraint engineering
- ☐ Teach someone else how constraints stimulate creative thinking

Week 4: Integrate and Apply

- ☐ Create your own systematic technique combining all three approaches

☐ Apply it to a meaningful challenge in your domain

☐ Document your process for future challenges

☐ Validate results through expert feedback or testing

☐ Develop decision criteria for when to use advanced techniques

Technique Selection Framework

The real advanced skill isn't just knowing these techniques—it's knowing when to use which approach based on problem type, available time, required accuracy, and your expertise level.

Decision Framework for Technique Selection

Use Recursive Chain of Thought When

- Problem requires layered analysis
- You need to examine assumptions systematically
- Time allows for multi-step process
- Accuracy requirements are moderate (not life-critical)

Use Meta-Prompting When

- Process improvement is the goal
- You have time for systematic iteration
- Quality can be measured objectively
- Learning the improvement process matters as much as the outcome

Use Constraint Engineering When

- Creative blocks need breaking
- Conventional approaches aren't working
- You want to explore unconventional solution spaces
- You have domain expertise to evaluate feasibility

Use Simple Prompting When

- Quick answers needed
- Straightforward problems
- High-stakes decisions requiring verified information
- You lack time for complex processes

Recognizing When Techniques Fail

Advanced techniques don't always work. Here's how to recognize ineffective outputs and recover:

Signs of Ineffective Advanced Prompting

- **Recursive Loops**: Reasoning becomes circular or contradictory
- **Meta-Prompting Stagnation**: Improvements plateau or become artificial
- **Constraint Confusion**: Solutions ignore practical requirements
- **Complexity for Complexity's Sake**: Process becomes more complex than the problem

Recovery Strategies

1. **Simplify the Approach**: Return to basic prompting methods
2. **Validate Independently**: Check key assumptions through external sources
3. **Seek Expert Input**: Consult domain experts for reality checks
4. **Test Incrementally**: Try solutions on a smaller scale before full implementation
5. **Know When to Stop**: Some problems are better solved through other means

The Bottom Line

These three advanced techniques—recursive chain of thought, meta-prompting loops, and constraint engineering—aren't magic solutions. They're systematic tools that can enhance your collaboration with AI when applied thoughtfully and with appropriate limitations in mind.

Master these, and you're not just using Claude anymore. You're collaborating with an AI system using sophisticated methods that can amplify your thinking when combined with human judgment, domain expertise, and practical validation. You're having conversations that leverage both human insight and AI pattern recognition—not because Claude is superior to human experts, but because these techniques create a structured form of human-AI collaboration when you need to explore complex problems systematically.

The prompting basics from Chapter 4, "The Art of Prompting: Getting Better Responses," got you started. These advanced strategies provide sophisticated tools for complex challenges. Use them wisely, use them with appropriate expectations, and prepare to enhance your problem-solving capabilities when combined with your human expertise and validation.

Ready to elevate your AI collaboration? Your next conversation with Claude will be more systematic, more thoughtful, and more likely to produce actionable insights—when you apply these techniques appropriately and validate the results.

Next up: Chapter 11, "Working with Claude's Special Features," explores Claude's special features—the hidden powers you didn't know existed. But first, practice these advanced techniques with realistic expectations. Start with something meaningful but not critical. Test the boundaries. See what happens when you prompt like an expert while thinking like a scientist.

CHAPTER 11

Working with Claude's Special Features

In This Chapter

- Discovering artifacts—your persistent workspace within Claude (note: feature availability varies by Claude version, platform, and subscription level)
- Mastering projects for long-term collaboration with manual context management
- Understanding extended thinking for complex problems
- Processing files like a document detective
- Managing context across marathon conversations

So far, we've been treating Claude like a really smart conversation partner. But here's the thing—Claude has superpowers hiding in plain sight. These aren't just neat tricks; they're game-changing features that transform Claude from a chatbot into a full-fledged work platform. It's like discovering your reliable Honda Civic has a hidden button that activates jet engines. Buckle up—we're about to explore the features that'll make you wonder how you ever lived without them.

Important Note Not all Claude implementations support all described features. Feature availability varies by Claude version, platform (web/mobile/API), and subscription level. Always verify current functionality rather than relying on static documentation as Claude features evolve rapidly.

R. Dickey, *Mastering Claude AI*, https://doi.org/10.1007/979-8-8688-2001-4_11

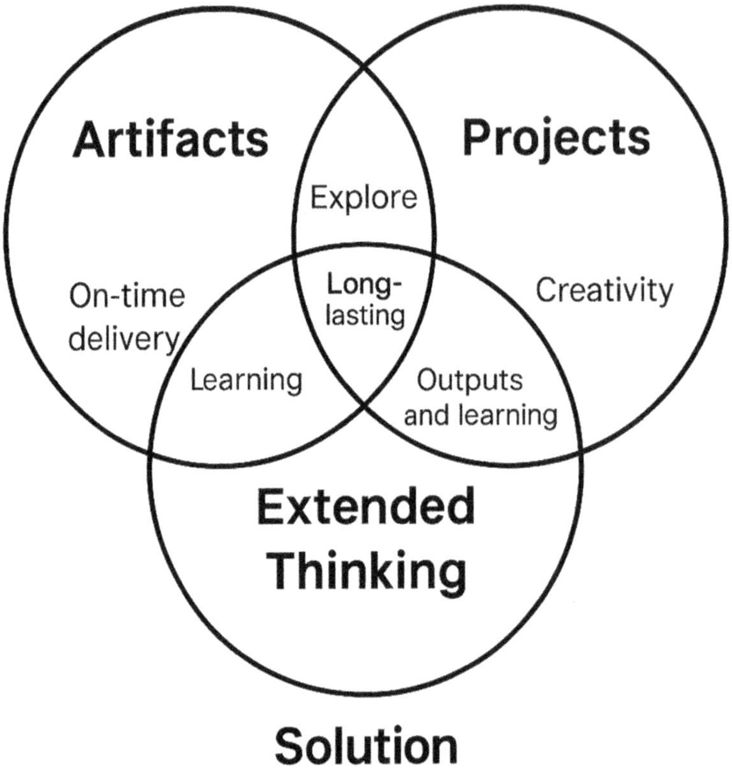

Figure 11-1. *The Claude features ecosystem: Artifacts, Projects, and Extended Thinking work together synergistically, with overlapping benefits that amplify productivity. Platform availability and feature access vary by subscription level and implementation*

Artifacts: Your Living Documents

Remember the frustration of asking Claude to write something, then losing it in the conversation scroll? Or worse, asking for edits and getting a completely new version? Enter Artifacts—Claude's answer to "I wish I could just edit that one part without starting over."

What Are Artifacts, Really?

Artifacts are persistent, editable documents that live alongside your conversation. Think of them as Google Docs, but built right into Claude. When Claude creates an artifact, it appears in a separate panel that you can

- Ask Claude to edit and update through conversation
- Copy with one click
- Keep visible while you discuss changes
- View across different versions

Technical Limitations to Know:

- File size limits vary by platform
- Supported formats include code, documents, and visualizations
- Processing constraints apply to complex content
- Platform-specific variations affect functionality
- Claude's performance may degrade in very long conversations due to technical limitations

It's the difference between having a conversation about a document and actually working on the document together through conversational edits.

When Claude Creates Artifacts

Claude doesn't create artifacts for everything (thank goodness—that would be chaos). Here's when they appear:

Automatic Artifact Creation:

- Substantial code (more than a quick snippet)
- Complete documents (emails, letters, reports)
- Creative writing (stories, scripts, poems)
- Structured content (meal plans, schedules, guides)
- Data visualizations (when possible)
- Technical documentation

No Artifact Needed:

- Quick answers or explanations
- Short responses under 20 lines

- Conversational replies
- Simple lists or calculations

Making the Most of Artifacts

The Living Document Approach

> Instead of "Write me a blog post about productivity."
>
> **Try**: "Create a blog post about productivity that I can refine with you. I'll want to adjust the tone and add personal examples as we go."

Claude creates an artifact you can request edits to while discussing changes in the chat.

The Iterative Development Method

> **You**: "I like the introduction, but can you make the second paragraph more conversational?"

Claude will update just that paragraph in the existing artifact. No more scrolling to find the latest version!

The Code Workshop

> **You**: "Create a Python script to organize my photo files by date."

Claude creates an artifact with the code. Then,

- "Add error handling for missing dates"
- "Include a progress bar"
- "Make it work with video files too"

Each request updates the same artifact. You watch your code evolve without losing track of changes.

Real-World Artifact Success Stories

The Newsletter Evolution (Illustrative Example) A marketing manager needed company newsletter content:

- **First Draft**: Claude created the artifact
- Through conversation, refined the financial section tone
- Updated specific sections without rewriting everything
- **Final Version**: True collaborative result

The Code Development Project (Illustrative Example) A developer needed a data analysis script:

- Started with basic functionality in an artifact
- Added features through conversation requests
- Fixed bugs without starting over
- **Final Script**: Iteratively developed through artifact updates

Artifact Pro Tips

Name Your Artifacts: "Create a project proposal artifact called 'Q4 Marketing Plan' that we can refine together"

Be Specific About Updates: "In the artifact, update only the budget section to reflect the new numbers"

Request Targeted Changes: Ask Claude for specific edits rather than direct modifications

Use Artifacts as Templates: "Create a meeting agenda artifact I can reuse weekly"

Projects: Your Persistent Workspace

If conversations with Claude are like meetings, then projects are like having a dedicated office where you can organize related work. Projects let you organize related conversations and store shared documents for reference across multiple sessions.

Understanding the Project Paradigm

Projects organize conversations like folders and allow files to serve as shared context across conversations within the project. However, they do not automatically maintain context between conversations.

Projects address three common organizational challenges:

1. **Scattered Related Work**: Projects organize conversations but do not eliminate the need to reestablish context in new conversations. Users must manually reference or summarize previous work.
2. **Lost Context Management**: Context maintenance requires manual effort—saving key information as artifacts or documents that can be referenced in future conversations within the project.
3. **Starting Fresh Each Time**: Projects do not automatically continue conversations. Users must manually provide context and reference previous work.

Think of a project as a smart folder that

- Organizes what you're working on
- Stores documents for reference across conversations
- Groups of related artifacts and discussions
- Requires manual context management for continuity

Creating Meaningful Projects

The Focused Project "Book Writing—Claude for Dummies"

- All chapter discussions in one place
- Consistent voice through saved style guidelines
- Character/concept continuity through reference documents
- Progress tracking is manual—users must document progress in artifacts or files within the project for future reference

The Ongoing Assistant Project "Weekly Business Analysis"

- Regular report formatting guidelines saved as reference
- Company context maintained through uploaded documents
- Previous insights available through saved artifacts
- Requires manual context setting for each new conversation

The Learning Journey Project "Learning Spanish with Claude"

- Progress tracking through saved lesson notes
- Weak spots documented in project files
- Previous lessons available for reference
- Consistent difficulty through manual context management

Context Management: The Manual Process

Context management requires active effort from users. Here's how to maintain continuity:

Smart Context Practices

Set Context Early*:* "This project is for planning my wedding. Key details from our project files: October 2025, 150 guests, outdoor venue, $30K budget, rustic theme."

Update Context As Needed: "Update: Based on our previous conversations saved in this project, we've decided on the venue—Willow Creek Farm. This changes our catering options."

Reference Previous Discussions*:* "Looking at our saved artifacts from last week's discussion about centerpieces, let's build on option #3."

Project Workflows That Work

The Serial Project Approach

1. Create a project for a major goal
2. Set up initial context documents
3. Save key decisions as artifacts
4. Reference previous work manually in new conversations
5. Build systematic documentation for continuity

The Document Library Method

1. Upload relevant background materials
2. Create reference artifacts for key decisions
3. Build conversation summaries for major discussions
4. Manually reference these materials in new chats

Project Success Framework

Systematic Documentation Strategy: Rather than relying on automatic context transfer, successful project users:

- Save key decisions in clearly named artifacts
- Create summary documents after important conversations
- Upload relevant background materials to project knowledge
- Start new conversations by referencing specific previous work
- Build a personal system for context continuity

Extended Thinking: When Claude Goes Deep

Sometimes your problems are too complex for quick responses. That's when Claude's extended thinking capabilities shine—allowing for more thorough analysis of complex challenges.

Triggering Extended Thinking

Complex prompts naturally encourage deeper analysis:

"Analyze the pros and cons of remote work for a 50-person startup, considering team dynamics, productivity metrics, cost implications, and long-term growth potential. Include implementation strategies for hybrid models."

Extended Thinking in Action

The Strategic Decision Matrix

Instead of quick answers, Claude can provide:

- Multi-dimensional analysis
- Consideration of competing factors
- Implementation pathways
- Risk assessment frameworks
- Long-term consequence evaluation

The Complex Problem Decomposition

"I need to decide whether to pivot our SaaS product. Consider market timing, technical debt, customer feedback, competitive landscape, and resource constraints."

Claude can work through systematic analysis covering each dimension thoroughly.

File Processing: Your Document Detective

Upload files to Claude and transform static documents into active insights. This feature turns Claude into your personal research assistant, document analyzer, and data interpreter.

What Files Can Claude Process?

Supported Formats:

- **Documents**: PDF, DOCX, TXT, RTF, ODT, HTML, EPUB
- **Data**: CSV, JSON

- **Images**: JPEG, PNG, GIF, WEBP (up to ~8,000 × 8,000 pixels)
- **Spreadsheets**: XLSX (with Analysis Tool enabled)

File Limitations:

- Maximum 20 files per conversation
- 30 MB size limit per file
- Text-focused processing for PDFs
- Platform-specific processing constraints

File Processing Strategies

The Research Synthesizer: "I've uploaded 5 research papers on renewable energy adoption. Can you

- Summarize key findings from each
- Identify common themes
- Note contradictions
- Suggest knowledge gaps
- Create a synthesis for my literature review"

The Document Analyzer: "Here's our 50-page contract. Please:

- Identify key terms and obligations
- Flag potential risks
- Summarize payment terms
- Note termination clauses
- Highlight anything unusual"

The Data Detective: "I've uploaded our sales data CSV. Can you:

- Identify trends
- Spot anomalies
- Compare regional performance

- Suggest focus areas
- Create a narrative for the board"

File Processing Best Practices

Prepare Your Files

- Use clear filenames
- Ensure text is searchable (not scanned images)
- Break massive files into logical chunks
- Remove sensitive information first
- Verify file format compatibility

  Guide the Analysis: Don't just upload and say "analyze this." Be specific: "Focus on Chapter 3's methodology." Set priorities: "I'm most interested in cost implications." Provide context: "This is for our annual review."

  Iterate on Insights: First pass: General analysis. Second pass: Dive into interesting findings. Third pass: Specific questions based on what you learned. Final pass: Actionable recommendations

File Processing Success Examples

The Contract Evaluation (Illustrative Example): A business owner uploaded a vendor contract and received an analysis of:

- One-sided termination clauses
- Automatic renewal mechanisms
- Pricing escalation issues
- Negotiation point recommendations
- Risk mitigation strategies

The Research Synthesis (Illustrative Example): A graduate student uploaded multiple academic articles and received:

- Theoretical framework mapping
- Methodological pattern identification
- Research gap analysis
- Comprehensive literature organization
- Structured review foundation

Managing Marathon Conversations

Long conversations with Claude can be incredibly productive—if you manage context well. Here's how to keep marathon sessions crisp and effective.

The Context Window Challenge

Remember, Claude has limits on how much it can "remember" at once. In long conversations,

- Earlier details may fade from active memory
- Context can get muddled
- Performance might degrade
- Responses may become generic

Context Management Strategies

The Checkpoint System: Every 10–15 exchanges: "Let's checkpoint. Here's what we've covered: [summary]. Here's what we're focusing on now: [current topic]. Here's what's next: [upcoming work]."

The Context Refresh: "We've been working for a while. Let me remind you of key context:

- **Project**: Website redesign
- **Target**: Young professionals
- **Constraints**: $10K budget, 6-week timeline
- **Current focus**: Homepage layout"

The Strategic Split: When conversations get too long:

- Save important outputs as artifacts
- Start a fresh conversation in the same project
- Provide a concise summary of previous work
- Continue with clean context

Marathon Session Success Tactics

The Progressive Build: Don't try to do everything at once

- **Session 1**: Planning and structure
- **Session 2**: First draft development
- **Session 3**: Refinement and polish
- **Session 4**: Final review and adjustments

The Context Journal: Keep a simple document with

- Key decisions made
- Important constraints
- Style/tone guidelines
- Progress milestones
- Next steps

Reference this at the start of each session.

Long Conversation Success Examples

The Novel Development Marathon (Illustrative Example): A writer spent multiple sessions developing a novel outline:

- Used checkpoints every chapter
- Maintained character consistency with context refreshes
- Split into multiple conversations when needed
- Result: Complete detailed chapter outline

The System Architecture Session (Illustrative Example): A developer designed a system architecture:

- Multiple sessions with clear handoffs
- Used artifacts for specifications
- Regular context management
- Clear summaries between sessions
- Result: Production-ready architecture documentation

Your Special Features Action Plan

Ready to level up with Claude's special features? Here's your roadmap:

Week 1: Master Artifacts

☐ Create your first artifact for a real project

☐ Practice requesting specific section updates

☐ Work with Claude to make iterative improvements

☐ Save an artifact as a reusable template

Week 2: Establish Projects

- ☐ Set up projects for your main work areas
- ☐ Practice manual context setting and management
- ☐ Create reference documents for continuity
- ☐ Experience the power of organized conversations

Week 3: Explore Deep Features

- ☐ Trigger extended thinking with complex problems
- ☐ Upload and analyze a real document
- ☐ Process data files for insights
- ☐ Compare manual vs. file processing efficiency

Week 4: Integrate Everything

- ☐ Run a marathon session with good context management
- ☐ Combine artifacts with file processing
- ☐ Use projects to organize related work
- ☐ Build your personal Claude workflow

Common Special Features Pitfalls

The Artifact Overuser: Don't request artifacts for everything. Quick responses don't need persistence.

The Context Assumer: Remember that context isn't automatically maintained between project conversations. Set context manually.

The File Dump: Don't upload files without clear analysis goals. Guide Claude's focus.

The Marathon Muddle: Take breaks. Refresh context. Split when needed. Quality over quantity.

The Automatic Context Expectation: Projects organize but don't automatically continue conversations. Manual effort maintains continuity.

The Bottom Line

These special features aren't just bells and whistles—they're what transform Claude from a smart chatbot into a true AI collaborator. Artifacts give you a shared workspace for iterative development. Projects organize your work with manual context management. Extended thinking tackles your toughest challenges. File processing unlocks your documents' insights. Context management keeps everything coherent.

Master these features, and you're not just chatting with Claude anymore. You're building a persistent, intelligent workspace where ideas evolve, projects progress, and complex problems get solved. It's the difference between having conversations and getting work done.

Remember: Start with one feature. Get comfortable. Understand the manual effort required. Then add another. Before you know it, you'll be orchestrating a symphony of AI capabilities that makes your old workflow look like banging rocks together.

Ready to unlock Claude's full potential? Your special features journey starts with your next conversation. Make it count—and this time, put it in a project so you can organize your progress systematically.

Next Up: Chapter 12, "Integration and Workflow Development," shows you how to integrate Claude into your daily workflow. But first, go create an artifact. Upload a file. Start a project. Feel the power of organized, persistent AI collaboration.

CHAPTER 12

Integration and Workflow Development

In This Chapter

- Building Claude into your daily routine without disrupting what works
- Creating workflows that actually make you more productive
- Establishing quality control systems that catch mistakes early
- Mastering collaboration between you, Claude, and your team (including technical requirements for shared access methods, consistent prompting strategies, and coordination mechanisms for multi-user workflows)
- Designing systems that grow with your success

You know that person who buys all the fancy organizing supplies but never actually gets organized? Or the one with seventeen productivity apps who's somehow less productive than before? Yeah, we're not doing that. This chapter is about making Claude a natural part of how you work, not another thing to manage. Think of it as the difference between bolting a GPS onto your dashboard vs. having it built into your car. The goal: Claude amplifying your work without adding a single extra step to your day.

R. Dickey, *Mastering Claude AI*, https://doi.org/10.1007/979-8-8688-2001-4_12

Integration: The Art of Seamless Enhancement

Integration—incorporating Claude into your existing workflows and tools—isn't about revolutionizing everything overnight. It's about finding the friction points in your day and smoothing them out, one at a time.

> **Important Considerations:** AI integration often introduces new complexities, including dependency on external services, data format requirements, and technical troubleshooting needs. Successful integration requires careful planning to address these challenges systematically.

Technical Infrastructure Prerequisites

Before diving into Claude integration, ensure your foundation can support AI-enhanced workflows:

Essential Technical Requirements:

- **Reliable Internet Connection:** AI services require consistent connectivity; intermittent connections disrupt workflows
- **Data Backup Systems:** Critical for protecting work when AI services experience outages
- **Alternative Workflow Plans:** Manual processes to maintain productivity during AI downtime
- **Security Framework:** Data handling protocols for information shared with AI services
- **Version Control:** Systems to track changes when AI assists with document creation

Integration Complexity Assessment:

- **System Compatibility:** Evaluate how AI tools connect with existing software
- **Data Format Requirements:** Ensure your data can be processed by AI systems

- **User Access Management:** Plan how team members will share AI resources
- **Cost Monitoring:** Track usage to avoid unexpected rate limits or billing issues

Start Where You Are

Sarah discovered this when she integrated Claude into her coffee shop operations:

> *"I don't want to change everything. I just want my current systems to work better."*

She started small:

- Morning inventory emails (Claude drafts, she personalizes)
- Daily special announcements (Claude suggests, she approves)
- Staff scheduling issues (Claude identifies conflicts; she resolves)
- Customer feedback analysis (Claude summarizes, she acts)

The key? Each integration made her existing process better, not different.

The Three Levels of Integration

Level 1: The Assistant Claude helps with specific tasks:

- Drafting emails you'll edit
- Researching topics you'll verify
- Creating first drafts you'll polish

Level 2: The Partner Claude becomes part of your thinking process:

- Brainstorming before big decisions
- Regular workflow check-ins
- Collaborative problem-solving

Level 3: The Amplifier Claude extends your capabilities:

- Taking on projects previously out of reach
- Working confidently in new domains
- Achieving quality levels is impossible alone

Most people find their sweet spot at Level 2—true **collaboration** where human judgment meets AI capability.

Workflow: Building Your Productivity Pipeline

A **workflow**—a sequence of steps to complete a task efficiently—is like a recipe for getting things done. With Claude, you can create workflows that feel less like chores and more like superpowers.

The Weekly Report Workflow

Marcus transformed his most dreaded task into a smooth process:

Trigger: Thursday 3 PM reminder

Step 1: Data Gathering (15 minutes)

- Marcus exports the week's data
- Uploads to Claude: "Analyze this week's marketing data for trends, anomalies, and notable changes"

Step 2: Insight Generation (20 minutes)

- Claude identifies key patterns
- Marcus adds context: "The spike on Tuesday was our flash sale"
- Together they find the real story

Step 3: Report Creation (10 minutes)

- Claude drafts based on insights
- Marcus reviews and personalizes
- Final polish together

Result: 45-minute process that used to take 2+ hours

Important Note Efficiency gains assume consistent AI performance. During high-demand periods, AI response times may vary, and occasional service interruptions can extend workflow completion times. Build buffer time into critical processes and maintain manual backup procedures.

Workflow Design Principles

Keep It Simple

- Start with 3–5 steps maximum
- Add complexity only when needed
- If it feels complicated, it is

Build In Checkpoints

- Review after each major step
- Catch errors early
- Maintain quality throughout

Leave Room for Humans

- Workflows guide; they don't dictate
- Keep the parts that need your touch
- Automate the repetitive, not the important

Plan for Variability

- Account for AI rate limits during peak usage
- Establish backup procedures for service outages
- Monitor performance and adjust expectations accordingly

Scalability and Performance Considerations

AI-Dependent Workflow Reality Check:

- Workflows may not scale linearly with increased usage
- Rate limits can create bottlenecks during high-demand periods
- Cost constraints may limit extensive AI utilization
- Performance can degrade with complex or high-volume requests

Success Metrics Framework:

- **Accuracy Rates:** Track error frequency and correction time
- **Time Savings Verification:** Measure actual vs. projected efficiency gains
- **Quality Maintenance:** Monitor output consistency over time
- **Total Cost Assessment:** Include AI service costs, training time, and maintenance overhead

Quality Control: Your Safety Net

Quality control—processes to ensure outputs meet standards—isn't about perfection. It's about catching the important stuff before it matters.

The Three-Layer Quality System

Dr. Jennifer Chen developed this for her research papers:

Layer 1: Claude's Self-Check: "Review this abstract for clarity, accuracy, and compliance with journal guidelines. Flag any concerns."

Layer 2: Human Judgment: Jennifer reviews for:

- **Technical accuracy** using specific criteria for detecting AI errors:
    - Factual inaccuracies requiring external verification
    - Logical inconsistencies within the content

    - Inappropriate tone for the intended audience
    - Output that doesn't match stated requirements

- **Appropriate tone**
- **Missing context**
- **Strategic decisions**

Layer 3: Final Polish: "Claude, help me polish this for publication. Check for consistency, grammar, and formatting."

This system caught errors that would have delayed publication by months.

Systematic Error Detection Framework

AI Error Categories to Monitor:

- **Factual Errors:** Information that contradicts verified sources
- **Logical Inconsistencies:** Contradictory statements within the same output
- **Tone Mismatches:** Inappropriate formality, urgency, or audience consideration
- **Requirement Deviations:** Output that ignores specified constraints or guidelines
- **Context Gaps:** Missing critical information for complete understanding

Validation Approaches:

- **Cross-Reference Verification:** Check claims against authoritative sources
- **Internal Consistency Review:** Scan for contradictory statements
- **Requirement Compliance:** Confirm output meets all specified criteria
- **Expert Review:** Involve domain specialists for technical accuracy

Quality Control Best Practices

For Writing:

- Claude checks grammar and structure
- You verify facts and tone
- Together ensure clarity

For Data:

- Claude identifies patterns and anomalies
- You validate conclusions against known benchmarks
- Both check for analytical biases

For Decisions:

- Claude provides an analysis framework
- You add contextual constraints and values
- Review assumptions together with domain expertise

Team Collaboration: Multi-user AI Integration

Technical Requirements for Team AI Integration:

Shared Access Management

- **Account Structure:** Establish organization-level AI service accounts vs. individual subscriptions
- **Usage Allocation:** Distribute rate limits and costs across team members
- **Permission Levels:** Define who can access sensitive AI features or data processing
- **Activity Monitoring:** Track individual usage for cost management and optimization

Consistent Prompting Strategies

- **Template Library:** Develop standardized prompts for common team tasks
- **Style Guidelines:** Establish tone and format standards for AI-generated content
- **Context Sharing:** Create systems for sharing project context across team AI interactions
- **Quality Standards:** Define acceptance criteria for AI-assisted outputs

Coordination Mechanisms for Multi-User Workflows

- **Handoff Protocols:** Procedures for transferring AI-assisted work between team members
- **Version Control:** Systems to track AI-assisted changes and collaborative editing
- **Conflict Resolution:** Methods for handling conflicting AI recommendations
- **Knowledge Management:** Centralized storage for AI interaction results and lessons learned

Productivity: It's Not About Speed

Productivity with Claude isn't about doing more things faster. It's about doing the right things better.

The Productivity Paradox

Tom discovered this truth:

> *"I was using Claude to do everything faster, but I wasn't getting more important work done."*

The shift came when he realized:

- Speed without direction is just busy
- Quality beats quantity every time
- Energy saved should go to what matters

His new approach:

- Claude handles routine tasks
- Tom focuses on strategic work
- Together they achieve more meaningful results

Productivity Metrics That Matter

Before Claude: Hours worked, tasks completed

With Claude: Impact achieved, problems solved

Track what actually moves the needle, not just what keeps you busy.
Realistic Productivity Expectations:

- Initial implementation may temporarily reduce efficiency during the learning curve
- Benefits compound over time as systems mature and teams adapt
- Individual results vary based on task complexity and AI integration depth
- Regular evaluation needed to ensure continued value delivery

Managing AI Capability Changes

Version Management and Update Impact Assessment:

Handling AI Evolution

- **Feature Deprecation:** Monitor announcements for discontinued capabilities
- **Performance Changes:** Track response quality and adjust expectations accordingly
- **Cost Structure Updates:** Evaluate the impact of pricing changes on workflow economics
- **New Capability Integration:** Assess whether new features improve existing workflows

Workflow Resilience Strategies

- **Version Documentation:** Record which AI capabilities specific workflows depend on
- **Backup Procedures:** Maintain manual alternatives for critical processes
- **Testing Protocols:** Regularly validate that workflows still function as expected
- **Adaptation Planning:** Prepare to modify workflows when AI capabilities change

Real-World Integration Success Examples

Sarah's Coffee Shop (Illustrative Example): Integrated Claude into daily operations, reduced admin time by approximately 50%, and spent more time with customers.

Marcus's Marketing (Illustrative Example): Built workflows for reports and campaigns, improved output quality while reducing work hours.

Dr. Chen's Research (Illustrative Example): Created quality control systems, streamlined revision cycles, and increased publication efficiency.

Tom's Consulting (Illustrative Example): Designed client workflows with Claude and improved service delivery efficiency.

Note Specific efficiency metrics represent individual experiences and may not reflect typical results. Actual benefits depend on implementation quality, user adaptation, and task compatibility with AI assistance.

Common Integration Pitfalls

The All-or-Nothing Trap: Trying to change everything at once. Start small, build gradually.

The Perfection Paralysis: Waiting for the perfect workflow. Good enough today beats perfect never.

The Tool Obsession: Focusing on features over outcomes. Remember why you're integrating.

The Human Replacement: Trying to automate relationship-building. Keep the human parts human.

The Dependency Assumption: Expecting AI to always be available. Plan for service interruptions and performance variability.

The Scaling Mirage: Assuming workflows will scale linearly. Monitor performance and costs as usage increases.

Your Integration Action Plan

Ready to make Claude part of your daily success? Here's your roadmap:

Week 1: Foundation and Assessment

- ☐ Evaluate technical infrastructure requirements
- ☐ List your three biggest time drains
- ☐ Choose one to tackle with Claude
- ☐ Create a simple 3-step workflow
- ☐ Test and refine for 5 days with performance monitoring

Week 2: Systematic Expansion

- ☐ Add quality control checkpoints with error detection criteria
- ☐ Integrate a second workflow
- ☐ Implement backup procedures for AI service interruptions
- ☐ Document what's working and measure actual efficiency gains

Week 3: Optimization and Team Integration

- ☐ Establish team coordination mechanisms if applicable
- ☐ Measure productivity improvements against baseline metrics
- ☐ Refine based on results and cost analysis
- ☐ Plan next integration targets based on proven value

Week 4: Resilience and Scaling

- ☐ Create version management procedures for AI capability changes
- ☐ Test backup workflows during simulated AI downtime
- ☐ Evaluate scaling limitations and cost implications
- ☐ Document lessons learned and best practices

The Bottom Line

Integration isn't about becoming dependent on AI or changing everything about how you work. It's about thoughtfully enhancing what you already do well while acknowledging the technical complexities and limitations involved. The best integration is one that improves your capabilities while maintaining resilience when AI services experience limitations.

Whether you're building **workflows** for daily tasks, establishing **quality control** systems, boosting **productivity**, or enhancing **collaboration**, the key is starting where you are and improving gradually with realistic expectations. Sarah still makes coffee with her hands, but Claude helps her run the business more efficiently. Marcus still creates marketing strategies, but Claude helps him execute them with greater consistency. That's the sweet spot—human creativity and judgment enhanced by AI capability, with appropriate backup plans and quality controls.

Remember: You're not trying to work like a robot. You're trying to work like a human with superpowers while maintaining the infrastructure and processes needed to succeed when those superpowers have limitations. Claude provides enhanced capabilities; you provide the humanity, strategic thinking, and resilience planning. Together, with proper preparation, that's powerful.

And with that, you've completed Part III! You've mastered advanced prompting strategies, discovered Claude's special features, and learned how to integrate everything into your daily workflow with realistic expectations and proper safeguards. You're no longer just using Claude—you're collaborating with it at an advanced level while maintaining professional standards and workflow resilience.

Ready for Part IV? We're about to explore how Claude transforms specific professional domains. Whether you're in business, education, or creative fields, the next section shows you how to apply everything you've learned to your specific world. The foundation is built with proper technical considerations. Now let's see what you can create on top of it.

PART IV

Professional and Specialized Applications

CHAPTER 13

Business and Professional Uses

In This Chapter

- Turning Claude into your business analysis assistant
- Creating professional documents that get read
- Conducting market research without breaking the bank
- Maintaining compliance without losing your mind
- Building systems that make your business run itself

Welcome to Part IV! You've mastered the fundamentals, explored practical applications, and learned advanced techniques. Now it's time to see how Claude transforms specific professional domains. Whether you're running a Fortune 500 company or a food truck, this section shows you how to apply everything you've learned to your professional world. Let's start with business, where Claude can be the difference between working harder and working smarter.

The biggest misconception about AI in business is that it requires a computer science degree and a massive IT budget. Reality check: Small businesses everywhere are using Claude to automate invoicing, improve customer service, analyze competition, and boost productivity. The only requirement is knowing what problems need to be solved.

Important Professional Considerations: Using AI for professional work involves specific risks and responsibilities. Different industries have varying standards for AI disclosure and professional liability. Some fields prohibit certain AI uses entirely, while

R. Dickey, *Mastering Claude AI*, https://doi.org/10.1007/979-8-8688-2001-4_13

others require specific compliance measures. Always check your professional guidelines, industry regulations, and organizational policies before implementing Claude in client work or sensitive business contexts. Consider consulting with legal counsel regarding liability protection and compliance requirements for your specific use case.

In this chapter, you discover how Claude can become a valuable business assistant—one who never needs coffee breaks and won't eat the last donut in the break room.

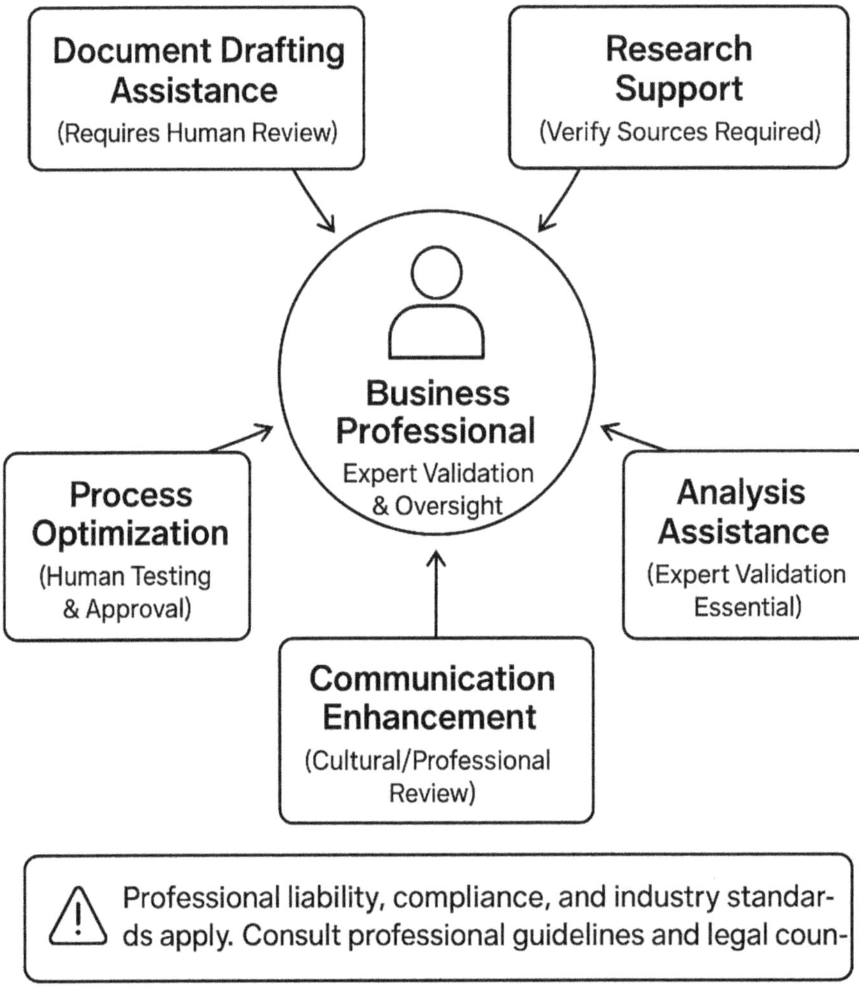

***Figure 13-1.** Claude at the center of business transformation: Five key areas where AI assistance revolutionizes professional work, from analysis to compliance*

Claude: Your 24/7 Business Consultant

Think of Claude as that expensive consultant you always wanted to hire, except Claude doesn't charge $500 an hour. Sarah discovered this when she needed help analyzing her coffee shop's sales data.

Understanding Business Intelligence with Claude's Assistance

Business Intelligence isn't just corporate speak—it's about making smart decisions based on data rather than gut feelings. Here's how Sarah used Claude as an analysis assistant:

> *"Claude, I have sales data from my coffee shop for Q3. Can you help me identify trends and suggest improvements? Here's the data: [sales figures]. Please look for patterns in product sales, peak hours, and day-of-week variations."*

Important Limitations: Claude can help interpret and summarize pre-calculated data but cannot perform complex statistical analysis, advanced calculations, or sophisticated business intelligence tasks independently. For best results, ensure your data is already summarized and calculations are completed using appropriate business tools before asking Claude for interpretation.

Claude helped Sarah identify that her pumpkin spice lattes outsold everything else on Thursdays and suggested optimizing staff schedules based on traffic patterns. However, Sarah verified these insights through her point-of-sale system's analytics before implementation. Results can vary significantly based on implementation quality, business context, and measurement methodology—specific outcomes cannot be guaranteed.

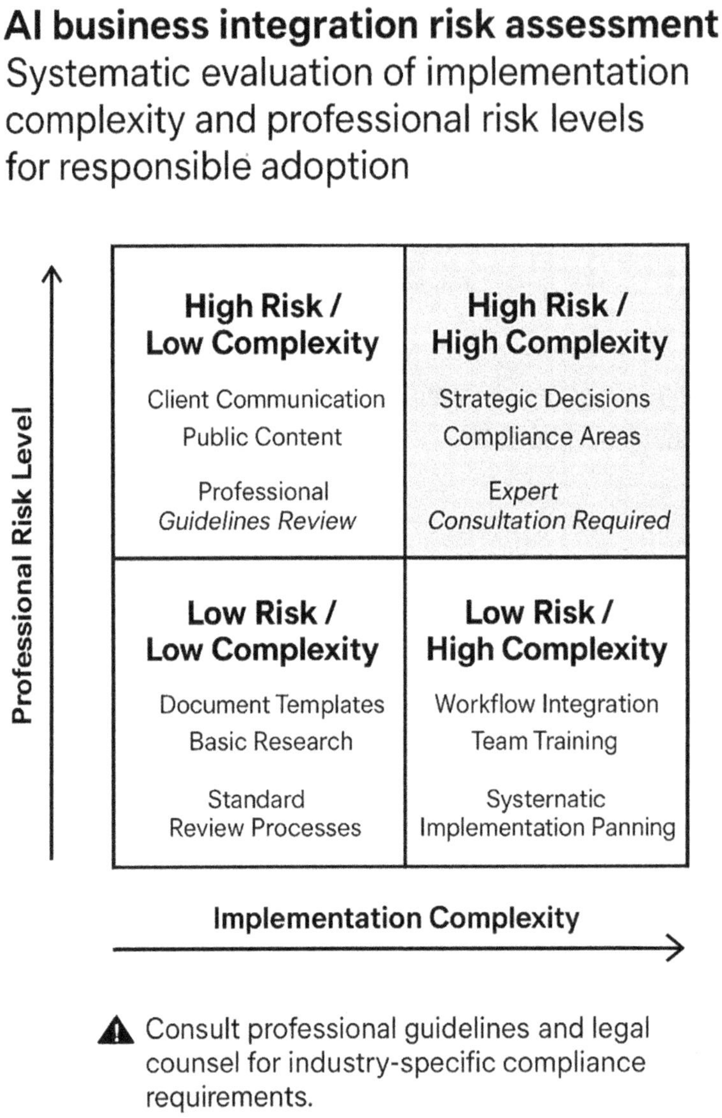

***Figure 13-2.** The ripple effect of Claude integration: Starting with efficiency at the core, benefits expand outward to quality, innovation, satisfaction, and ultimately revenue growth*

Professional Standards Framework

Professional Standards are your secret weapon for standing out. Claude can help you maintain these standards without turning into a corporate robot by following this systematic approach:

> **Step 1: Assessment:** Define your professional requirements and industry standards
>
> **Step 2: Content Development:** Use Claude to draft materials meeting those standards
>
> **Step 3: Compliance Review:** Verify adherence to professional guidelines
>
> **Step 4: Quality Control:** Implement human oversight and approval processes

Marcus learned this preparing a presentation for his marketing internship:

> *"I need to create a marketing presentation about social media trends for a corporate audience. Can you help me structure it professionally while keeping it engaging? Please note any areas where I should verify current data independently."*

Claude helped him create a presentation framework that was both professional and personality-filled, but Marcus verified all statistics and claims through current sources. His supervisor was impressed with the structure and asked him to present to the executive team!

Market Research with Appropriate Limitations

Competitive Analysis used to mean hiring expensive firms. Now Claude can help you research systematically, with proper understanding of its limitations.

The Systematic Research Framework

Step 1: Research Planning: Define specific research questions and methodology

Step 2: Data Collection: Gather information from validated sources

Step 3: Claude-Assisted Analysis: Use Claude to help interpret and summarize findings

Step 4: Expert Validation: Verify insights through domain expertise and additional sources

Tom discovered Claude's research assistance capabilities when starting his freelance technical writing business:

> *"Claude, I'm starting a technical writing business focused on software documentation. Based on the industry data I provide, can you help me analyze what successful technical writers typically charge, what services they offer, and how they position themselves? Please indicate where I should seek additional validation."*

Technical Requirements: Effective market research requires domain expertise, validated methodologies, and current market data that Claude cannot provide independently. Claude can help synthesize and interpret information you provide but cannot replace specialized market research tools or professional analysis.

Claude helped analyze the patterns Tom found in his research, but he verified pricing and market conditions through professional networks and industry surveys. The combination approach helped him identify niche specializations and pricing structures more efficiently than either method alone.

Creating Your Competitive Edge with Strategic Analysis

When a Starbucks opened across from Sarah's coffee shop:

> *"Claude, help me identify unique value propositions and marketing strategies that highlight what makes small, local coffee shops special. I need to compete on more than just price. Please help me structure my analysis of the competitive landscape I've researched."*

Sarah launched a "Know Your Barista" campaign emphasizing personal connections, but only after validating the approach through customer surveys and local market analysis. Her sales actually increased after the chain opened!

Document Creation That Gets Results

Professional doesn't have to mean boring. Claude can help you strike the perfect balance using our systematic framework.

Professional Document Framework

Step 1: Purpose Definition: Clearly define document objectives and audience

Step 2: Structure Development: Create logical flow with Claude's assistance

Step 3: Content Creation: Draft with professional standards in mind

Step 4: Review and Refinement: Implement quality control and approval processes

Proposals That Win

Dr. Jennifer Chen needed a research grant proposal:

> *"Claude, help me write a grant proposal for my autism research project. Make it comprehensive and professional while keeping it engaging enough that reviewers actually want to fund it. Please note any areas requiring specialized expertise or current research validation."*

Implementation Template:
"Claude, help me structure a [document type] for [specific audience].

Requirements:

- Professional tone for [industry/context]
- Include sections for [specific elements]
- Maintain engagement while meeting [specific standards]
- Flag areas requiring expert review or current data verification

Please provide the framework first, then we'll develop the content systematically."

Claude helped create a proposal structure that was rigorous yet readable, but Dr. Chen worked with grant writing specialists to ensure compliance with specific funding requirements. She got the grant on her first try!

Reports People Actually Read

"Claude, I need to write a quarterly business report for my investors. Structure it so it's professional but engaging. Make it something they'll actually want to read, while ensuring all financial data and claims are properly sourced."

The key is balancing professionalism with personality—something Claude excels at helping you achieve with proper oversight.

Compliance Without Complications

Compliance doesn't have to feel like navigating a minefield. Claude can help break down complex regulations into actionable steps, with appropriate verification requirements.

Compliance Management Framework

Step 1: Regulatory Research: Identify applicable regulations for your industry

Step 2: Claude-Assisted Interpretation: Use Claude to help understand requirements in plain language

Step 3: Professional Validation: Verify interpretations with legal or compliance experts

Step 4: Implementation Planning: Create systematic compliance processes

Marcus needed GDPR compliance for email campaigns:

"Claude, can you help me understand GDPR requirements for email marketing in plain English? Focus on practical steps we need to take. Please indicate where I should seek legal review for my specific situation."

Implementation Template:

"Claude, help me create a compliance checklist for [specific regulation] in [industry context].

Requirements:

- Plain language explanations
- Practical implementation steps
- Areas requiring legal review
- Regular update procedures

Please structure this as an actionable framework rather than legal advice."

Claude broke down regulations into actionable steps, complete with examples and template language, but Marcus had the final compliance plan reviewed by legal counsel familiar with current GDPR enforcement. What could have been weeks of confusion became a structured afternoon of implementation planning.

Sarah used this approach for health department regulations:

"Claude, help me create a daily checklist for my coffee shop that ensures we meet all health requirements. Make it simple enough for new employees, and indicate where I should verify current local regulations."

Now her staff follows easy checklists that keep inspectors happy without complexity, updated regularly based on current local requirements.

Professional Risk Management Framework

Critical Professional Considerations for AI use in business contexts:

Risk Assessment Procedures

Step 1: Liability Analysis: Assess professional liability implications for your industry

Step 2: Confidentiality Review: Evaluate client data and privacy requirements

Step 3: Compliance Verification: Check industry-specific AI usage regulations

Step 4: Documentation Standards: Establish proper record-keeping for AI assistance

Industry-Specific Guidelines

- **Legal Professionals**: Check bar association guidelines on AI disclosure requirements
- **Healthcare**: Verify HIPAA compliance and medical liability standards
- **Financial Services**: Review SEC, FINRA, or relevant regulatory guidance
- **Consulting**: Establish client disclosure policies for AI-assisted work
- **Government Contractors**: Verify security clearance and data handling requirements

Implementation Safeguards

Professional AI Use Checklist:

☐ Review professional liability insurance coverage for AI-assisted work

☐ Establish client disclosure policies about AI assistance

☐ Create backup processes for AI failure scenarios

☐ Implement human oversight for all client-facing deliverables

☐ Verify compliance with industry-specific regulations

☐ Document AI usage for audit and review purposes

☐ Establish data security protocols for sensitive information

Real-World Success Stories

The Overwhelmed Accountant: Used Claude to create client onboarding document templates, reducing setup time from 3 hours to 45 minutes per client while maintaining professional standards through systematic review processes.

The Expanding Retailer: Used Claude to help analyze customer feedback patterns, discovered 67% wanted eco-friendly packaging, then validated findings through additional customer surveys before implementing a brand transformation that doubled sales.

The Confused Consultant: Used Claude to structure service packages and pricing frameworks, going from uncertainty to $10K/month in recurring revenue after validating approaches with industry mentors.

Common Business Pitfalls to Avoid

The Automation Addiction: Don't automate everything. Personal touch still matters, especially in professional relationships.

The Data Dump: Focus on actionable insights, not overwhelming analysis. Remember Claude's limitations with complex calculations.

The Perfectionist Paralysis: Good enough today beats perfect never, but maintain professional standards.

The Compliance Shortcut: Never rely solely on Claude for legal or regulatory compliance—always verify with qualified professionals.

Your Business Transformation Checklist:

Ready to revolutionize your business? Here's your systematic action plan with proper risk management:

Week 1: Assessment and Risk Analysis

- ☐ Analyze your current business challenges with Claude's assistance
- ☐ Research your top 3 competitors using validated sources
- ☐ Identify your unique value proposition with Claude's help
- ☐ List repetitive tasks that could be systematized
- ☐ Assess professional liability and compliance requirements
- ☐ Identify potential AI failure points and backup processes
- ☐ Review industry-specific regulations for AI use

Week 2: Implementation with Safeguards

- ☐ Draft or improve one key business document with Claude's assistance
- ☐ Create standard operating procedures for one process
- ☐ Develop templates for common communications
- ☐ Build a compliance checklist for your industry

☐ Implement human review processes for all AI-assisted work

☐ Establish client disclosure policies where required

☐ Create documentation standards for AI usage

Week 3: Growth and Validation

☐ Create a 90-day business plan framework with Claude's help

☐ Implement at least 3 Claude-assisted improvements with proper oversight

☐ Track results and refine your approach systematically

☐ Plan your next month of Claude-powered growth

☐ Verify all outcomes through independent validation

☐ Review and update risk management procedures

☐ Conduct stakeholder feedback sessions on AI integration

Technical Implementation Framework:

Data Requirements and Processing Limitations

Understanding Claude's Technical Constraints:

- Input size limitations for data analysis
- Cannot perform complex statistical calculations independently
- Requires pre-processed, summarized data for optimal results
- Cannot access real-time databases or business systems directly

Integration Challenges and Solutions

Systematic Integration Approach:

Step 1: Technical Assessment

- Evaluate your current business systems and data infrastructure
- Identify integration points where Claude can add value
- Assess technical limitations and workaround requirements

Step 2: Pilot Implementation

- Start with low-risk, non-critical business processes
- Establish clear success metrics and evaluation criteria
- Create feedback loops for continuous improvement

Step 3: Scaling Strategy

- Develop systematic processes for broader implementation
- Establish training protocols for team members
- Create maintenance and update procedures

Quality Control Framework

Implementation Template for Technical Quality Control:

"Claude, help me review this [business analysis/document/process] for:

- Accuracy of interpretations
- Completeness of analysis
- Areas requiring additional verification
- Potential limitations or blind spots

Please flag any claims that require independent validation."

The Bottom Line

Using Claude for business isn't about replacing human intelligence—it's about amplifying it strategically while maintaining professional standards and managing appropriate risks. Think of Claude as your business's Swiss Army knife: versatile, reliable, and always ready to help, but requiring skilled operation and proper maintenance.

Sarah's coffee shop is thriving, Marcus landed a full-time position, Dr. Chen's research is funded, and Tom's writing business is booked solid. They succeeded because they learned to use Claude strategically while maintaining professional standards and managing risks appropriately.

Whether you're conducting **Business Intelligence Analysis** (with proper validation), maintaining **Professional Standards** (with appropriate oversight), conducting **Competitive Analysis** (with expert verification), or ensuring **Compliance** (with legal review), Claude can be your tireless business assistant. Your competition is probably still doing things the old way. Meanwhile, you have a 24/7 business advisor ready to help you leap ahead—with proper safeguards in place.

Final Framework Summary:

1. **Assess** your professional requirements and risk factors
2. **Implement** Claude assistance with appropriate limitations and oversight
3. **Validate** all outputs through qualified experts and independent verification
4. **Monitor** ongoing compliance and professional standards
5. **Refine** your approach based on results and changing requirements

Remember: Every business empire started with someone deciding to do things differently while maintaining the highest professional standards. Today, that someone is you. Let's get to work—responsibly and professionally!

CHAPTER 14

Education and Learning Applications

In This Chapter

- Transforming how you learn with AI-powered techniques
- Creating study guides that actually help you remember
- Developing new skills without expensive courses
- Maintaining academic integrity while leveraging AI
- Building personalized learning paths for any subject

Important Learning Effectiveness Disclaimer: Learning outcomes vary significantly by individual, subject matter, and educational context. Success stories presented are anecdotal examples, not guaranteed results. AI-enhanced learning effectiveness depends on multiple factors, including learning style, motivation, prior knowledge, implementation quality, and educational support systems. Individual results may differ substantially.

These success stories illustrate potential applications: Marcus aced his data analytics final while working full-time. Tom learned Python at 67 and now automates his entire home budget. Sarah's daughter went from failing AP History to getting the highest score in her class. What do they have in common? They discovered that Claude can help serve as a patient learning assistant, available 24/7, who explains things in multiple ways until they find what works for their brain. No judgment. No rush. Just learning support that adapts to individual needs.

R. Dickey, *Mastering Claude AI*, https://doi.org/10.1007/979-8-8688-2001-4_14

Learning Outcomes: What Success Actually Looks Like

Learning outcomes—the specific skills or knowledge gained from educational experiences—used to be something professors wrote on syllabi that nobody read. With Claude's assistance, they become your roadmap to actual understanding.

Defining Your Real Goals with a Systematic Framework

> **Step 1: Objective Translation**: Convert vague learning goals into specific, measurable abilities
>
> **Step 2: Skill Mapping**: Identify component skills needed for mastery
>
> **Step 3: Application Context**: Connect learning to real-world usage
>
> **Step 4: Progress Validation**: Establish checkpoints for verification through independent assessment

Marcus discovered this difference preparing for his data analytics course:

> "Claude, I need to 'understand statistical analysis' according to my syllabus. But what should I actually be able to DO by the end of this unit?"

Claude helped him translate vague objectives into concrete abilities:

- Calculate and interpret mean, median, and standard deviation
- Identify when each measure is most appropriate
- Spot misleading statistics in real-world examples
- Apply statistical thinking to marketing decisions

Now he wasn't just memorizing formulas—he was building skills he'd actually use, verified through practice problems and instructor feedback.

The Outcome-First Approach

Before diving into any learning:

> "Claude, I want to learn [topic]. What are the 5 most important things I should be able to do/explain/create when I'm done? Please help me structure a progression that I can validate through [expert review/formal assessment/practical application]."

This flips traditional learning on its head. Instead of wading through content hoping something sticks, you're targeting specific abilities from day one with clear validation pathways.

Skill Development: From Zero to Competent

Skill development—acquiring new abilities or improving existing ones—isn't about talent. It's about smart practice with appropriate validation. Claude can help make that practice smarter.

The Progressive Skill Building Framework

> **Step 1: Foundation Assessment**: Evaluate current skill level and learning requirements
>
> **Step 2: Structured Progression**: Create a systematic learning pathway with clear milestones
>
> **Step 3: Practice Implementation**: Apply skills through guided exercises and real projects
>
> **Step 4: Expert Validation**: Verify competency through qualified assessment beyond AI interaction

Critical Competency Development Requirement: Competency development requires practical application, feedback from qualified experts, and validation beyond AI interaction. Self-assessment with AI assistance is insufficient for skill verification. Always seek independent confirmation of skill development through qualified instructors, professional mentors, or industry-standard assessments.

Tom wanted to learn Python at age 67:

> "Claude, I've never programmed before. Help me create a 30-day skill development plan for Python that builds from absolute zero. Focus on practical projects I can actually use, and indicate where I should seek human feedback or formal validation."

Claude helped create a progression structure:

- **Days 1-5**: Basic syntax through interactive examples (validated through coding exercises)
- **Days 6-10**: First mini-project (expense tracker) with peer review
- **Days 11-15**: Working with files and data (expert feedback session)
- **Days 16-20**: Automating a daily task (mentor evaluation)
- **Days 21-25**: Building something useful (community code review)
- **Days 26-30**: Polishing and expanding (instructor assessment)

Each day built on the last with human validation at key milestones. No overwhelming leaps, but genuine skill verification.

Skills That Stick with Validation

The key to lasting skill development:

> **Practice with Purpose**: "Claude, give me a Python challenge that uses what I learned yesterday but adds one new concept. Then I'll have my mentor review my solution."
>
> **Immediate Application**: "How can I use this Excel function to solve a real problem in my life today? I'll verify the results with my colleague."
>
> **Spaced Repetition**: "Create a quick review quiz of Python concepts I learned this week, focusing on what I struggled with. I'll discuss the results with my study group."

Study Guide Creation: Your Personal Learning Assistant

A **study guide** isn't just a summary—it's your personalized learning tool that adapts to how YOUR brain works, with appropriate content validation.

The Adaptive Study Guide System with Validation Framework

> **Step 1: Content Analysis**: Identify key concepts and learning objectives
>
> **Step 2: Claude-Assisted Organization**: Structure materials for your learning style
>
> **Step 3: Expert Verification**: Validate content accuracy through authoritative sources
>
> **Step 4: Iterative Refinement**: Update based on learning effectiveness and expert feedback

Sarah's daughter was struggling with AP History:

> "Claude, here's the chapter on the Industrial Revolution. Create a study guide that:
>
> - Identifies the five most important concepts
> - Explains each in simple terms
> - Provides memory tricks for dates/names
> - Includes practice questions
> - Connects to modern examples she'd understand
> - Indicates where I should verify information with her teacher or textbook"

The result wasn't just notes—it was a learning experience tailored to her needs with clear verification pathways.

Study Guide Templates with Content Validation

For Conceptual Subjects: "Create a study guide for [topic] that includes:

- Core concepts in plain English
- Visual analogies for complex ideas
- Common misconceptions to avoid
- Practice problems with solutions
- Real-world applications"

Critical Content Warning: AI-generated study materials may contain factual errors or outdated information, particularly in rapidly evolving fields. All content requires verification against authoritative sources, including current textbooks, peer-reviewed materials, instructor guidance, and established educational standards, before use in academic work.

For Fact-Heavy Subjects: "Build a study guide for [topic] with:

- Key facts organized by theme
- Memory devices (mnemonics, stories)
- Timeline of important events
- Quick-reference charts
- Self-test questions
- Source verification checklist for fact-checking"

Systematic Content Validation Framework:

1. **Expert Review**: Have a qualified instructor or subject matter expert review all content
2. **Source Verification**: Cross-reference facts against multiple authoritative sources
3. **Standards Alignment**: Ensure alignment with current educational standards and curriculum

4. **Currency Check**: Verify information is current and reflects the latest developments
5. **Bias Assessment**: Evaluate potential biases or missing perspectives in content

Academic Integrity: Using AI the Right Way

Academic integrity—honest and ethical scholarship—isn't about avoiding AI. It's about using it appropriately to enhance genuine learning while meeting institutional requirements.

Critical Academic Integrity Considerations

Institutional Policy Variation: Academic integrity policies regarding AI use vary dramatically across educational levels and institutions. Elementary schools, high schools, colleges, and universities each have different standards. Individual instructors may have specific course policies that override general institutional guidelines. Always check:

- Your institution's official AI use policy
- Individual course syllabi and instructor guidelines
- Professional program requirements (medical, legal, engineering)
- Regional or accreditation body standards
- International study program policies

The Right Way to Use Claude for School

Marcus learned this balance through systematic guidance:

Generally Appropriate Uses (verify with your institution):

- "Explain this concept I don't understand"
- "Help me find gaps in my argument"
- "Suggest sources I should research"

- "Check my understanding with practice problems"
- "Help me outline my thoughts"

Generally Inappropriate Uses (prohibited by most institutions):

- "Write my essay for me"
- "Do my homework"
- "Take my online test"
- "Create something I'll submit as my own"

The Learning Enhancement Framework

Step 1: Policy Verification: Confirm institutional and course-specific AI policies

Step 2: Purpose Definition: Clearly define learning vs. completion objectives

Step 3: Attribution Planning: Determine disclosure and citation requirements

Step 4: Validation Process: Ensure your understanding, not AI dependence

Dr. Chen teaches her students:

"Claude can serve as your study partner, not your ghostwriter. Use it to:

- Understand difficult concepts
- Practice and get feedback
- Organize your own ideas
- Improve your writing process
- Prepare for discussions

But the work you submit must reflect YOUR understanding, YOUR analysis, and YOUR critical thinking. Important: Always disclose AI assistance per course policies."

Critical AI Dependence Warning: Over-dependence on AI for learning can lead to reduced critical thinking development, diminished problem-solving skills, and inability to function without AI assistance. While AI can enhance learning, excessive reliance may undermine the fundamental cognitive skills education aims to develop. Maintain balance between AI assistance and independent thinking.

Age-Appropriate AI Guidelines and Supervision

For Younger Learners (K-12): Educational AI use requires careful consideration of developmental appropriateness and supervision requirements:

- **Elementary (K-5)**: Direct adult supervision required for all AI interactions
- **Middle School (6-8)**: Guided use with regular teacher/parent oversight
- **High School (9-12)**: Increased independence with clear boundaries and periodic check-ins

Supervision Framework for Younger Learners:

1. **Adult Pre-Screening**: Review AI outputs before student interaction
2. **Learning Objective Alignment**: Ensure AI use supports educational goals
3. **Critical Thinking Protection**: Maintain opportunities for independent reasoning
4. **Digital Citizenship**: Teach responsible AI interaction and limitations
5. **Privacy Protection**: Ensure age-appropriate data handling and privacy practices

Parental or instructor oversight is particularly important for ensuring AI interactions support rather than replace critical thinking development.

Real-World Learning Success with Validation

Systematic Approaches for Validating AI-Generated Educational Content:

Expert Review Process:

- Qualified subject matter experts review all AI-generated materials
- Cross-reference with current academic standards and curriculum requirements
- Validate learning progression and pedagogical appropriateness

Source Verification Methods:

- Multiple authoritative source cross-checking
- Peer-reviewed material confirmation
- Current edition textbook alignment
- Professional organization standard compliance

Learning Standards Alignment:

- State and national educational standard verification
- Accreditation body requirement confirmation
- Professional certification standard alignment
- International education standard compatibility

The Career Changer: Marcus used Claude to learn data analysis concepts, supplementing his marketing degree. Combined AI assistance with online courses, mentor feedback, and professional certification. Got promoted within 6 months after demonstrating validated competency.

The Lifelong Learner: Tom mastered Python fundamentals at 67, using Claude for concept explanation alongside formal online coursework and coding community participation. Now he automates his home budget system and teaches others through community college workshops.

The Struggling Student: Sarah's daughter improved from C's to A's in AP History by creating personalized study guides with Claude's assistance, combined with teacher review, study group participation, and regular instructor office hours. Age-appropriate supervision ensured proper academic support rather than dependence.

The Fast Tracker: Dr. Chen's graduate students use Claude to understand complex papers more efficiently, combined with faculty discussion sessions, peer review groups, and structured research mentorship. This leaves more time for original research under expert guidance.

Learning Strategies That Actually Work

The Feynman Technique with Claude and Validation

"Claude, I just learned about [concept]. Let me explain it to you as if you're a 10-year-old. Tell me what parts of my explanation are unclear or incorrect. Then I'll verify my understanding with [instructor/study group/expert source]."

This forces you to truly understand, not just memorize, with external validation of comprehension.

The Problem-First Method with Expert Feedback

Instead of reading and then practicing: "Claude, give me a problem involving [concept]. Let me try to solve it, then help me understand where I got stuck. I'll discuss my solution approach with my instructor to confirm understanding."

Learning sticks better when you see why you need it and confirm your reasoning through qualified feedback.

The Connection Builder with Source Verification

"Claude, I just learned about [new concept]. How does this connect to [something I already know]? What are the similarities and differences? Please suggest authoritative sources where I can verify these connections."

Building connections between ideas creates lasting understanding when verified through reliable sources.

Common Learning Pitfalls

The Information Hoarder: Collecting resources without studying. Focus on understanding a few things deeply with proper validation.

The Passive Reader: Reading without engaging. Always interact with Claude about what you're learning, then verify through other sources.

The Perfectionist Procrastinator: Waiting to feel "ready." Start messy; improve through practice and expert feedback.

The Academic Integrity Violator: Using Claude to avoid learning. Remember: the goal is understanding, not just grades. Check institutional policies.

The AI Dependence Trap: Relying too heavily on AI assistance instead of developing independent critical thinking skills.

Your Learning Transformation Checklist:

Ready to revolutionize how you learn? Here's your systematic action plan with proper validation:

Week 1: Foundation Building with Policy Verification

- ☐ Define clear learning outcomes for one subject
- ☐ Review your institution's AI use policies and course-specific guidelines
- ☐ Create your first adaptive study guide with content verification
- ☐ Try the Feynman Technique with one concept and expert validation
- ☐ Set up a skill development plan with milestone checkpoints
- ☐ Identify qualified experts or instructors for ongoing feedback

Week 2: Active Practice with Validation

☐ Use the problem-first method daily with solution verification

☐ Build connections between new and old knowledge through multiple sources

☐ Create practice tests with Claude and review with instructor

☐ Track which methods work best for you through documented assessment

☐ Implement age-appropriate supervision if working with younger learners

☐ Establish systematic content verification processes

Week 3: Integration and Mastery with Expert Confirmation

☐ Apply new skills to real projects with mentor review

☐ Teach someone else what you've learned to test comprehension

☐ Refine your learning system based on feedback and results

☐ Plan your next learning adventure with realistic expectations

☐ Validate competency through qualified assessment beyond AI interaction

☐ Document learning outcomes and areas requiring continued development

The Bottom Line

Learning with Claude isn't about shortcuts or cheating—it's about learning smarter, faster, and deeper through systematic collaboration with appropriate validation. Whether you're developing **skills** for career advancement (with expert verification), creating **study guides** for tough subjects (with content validation), defining clear

learning outcomes (with measurable assessment), or maintaining **academic integrity** (with institutional compliance) while leveraging AI, the key is using Claude as a learning amplifier with proper human oversight and verification.

Critical Success Framework:

1. **Policy Compliance**: Always follow institutional guidelines and course requirements
2. **Content Validation**: Verify all AI-generated educational content through authoritative sources
3. **Expert Engagement**: Maintain regular interaction with qualified instructors and mentors
4. **Independent Thinking**: Preserve and develop critical thinking skills alongside AI assistance
5. **Appropriate Supervision**: Ensure age-appropriate oversight for younger learners
6. **Competency Verification**: Validate learning through qualified assessment beyond AI interaction

Marcus is acing courses while working full-time through systematic learning with validation. Tom is coding at 67 with proper mentorship and community support. Sarah's daughter enjoys studying with appropriate academic support. Dr. Chen's students understand complex topics efficiently while maintaining research integrity. They succeeded by learning how to learn with Claude while maintaining educational best practices and expert validation.

Remember: Your brain + Claude's capabilities + proper human guidance = enhanced learning potential. The question isn't whether you can learn something new. It's what you want to learn first, and how you'll validate that learning through appropriate expert assessment and real-world application.

Individual learning outcomes depend on multiple factors, including motivation, prior knowledge, learning style, implementation quality, and educational context. AI-enhanced learning requires careful integration with established educational practices and expert oversight to achieve optimal results while maintaining academic integrity and developing essential critical thinking skills.

CHAPTER 15

Creative and Artistic Collaboration

In This Chapter

- Unleashing your creative potential with an AI assistant
- Transforming writer's block into writer's flow
- Exploring artistic development across all media
- Creating content that captures hearts and minds
- Navigating intellectual property in the AI age

Critical Technical Understanding: How AI Creative Assistance Actually Works

Before exploring creative collaboration, it's essential to understand what AI can and cannot do creatively. Claude provides pattern-based creative suggestions from its training data, not genuine creative understanding or consciousness. It cannot truly evaluate artistic merit, provide authentic creative critique, or understand cultural context the way humans do. While AI can assist with creative processes through pattern recognition and combinatorial possibilities, breakthrough innovation, authentic artistic expression, and meaningful cultural creation require human insight, emotion, domain expertise, and real-world understanding that AI cannot provide.

Important Communication Note: This chapter uses practical language like "Claude sees" or "Claude becomes your creative partner" for ease of communication. These are metaphorical descriptions of pattern-based responses, not literal consciousness or thinking. Claude processes language patterns and provides responses based on training data, rather than having genuine thoughts, emotions, or creative experiences.

R. Dickey, *Mastering Claude AI*, https://doi.org/10.1007/979-8-8688-2001-4_15

Street artists use Claude to name installations. Composers break through creative blocks by exploring AI-generated variations. Novelists discover new directions by using Claude to examine plot possibilities. The creative world isn't being replaced by AI—it's finding new ways to enhance human creativity through systematic collaboration. Whether you paint, write, compose, design, perform, or create in ways that don't have names yet, Claude can help serve as a creative thinking assistant. One who never dismisses ideas, is always available for brainstorming sessions, and occasionally suggests combinations so unexpected they spark genuine breakthrough thinking.

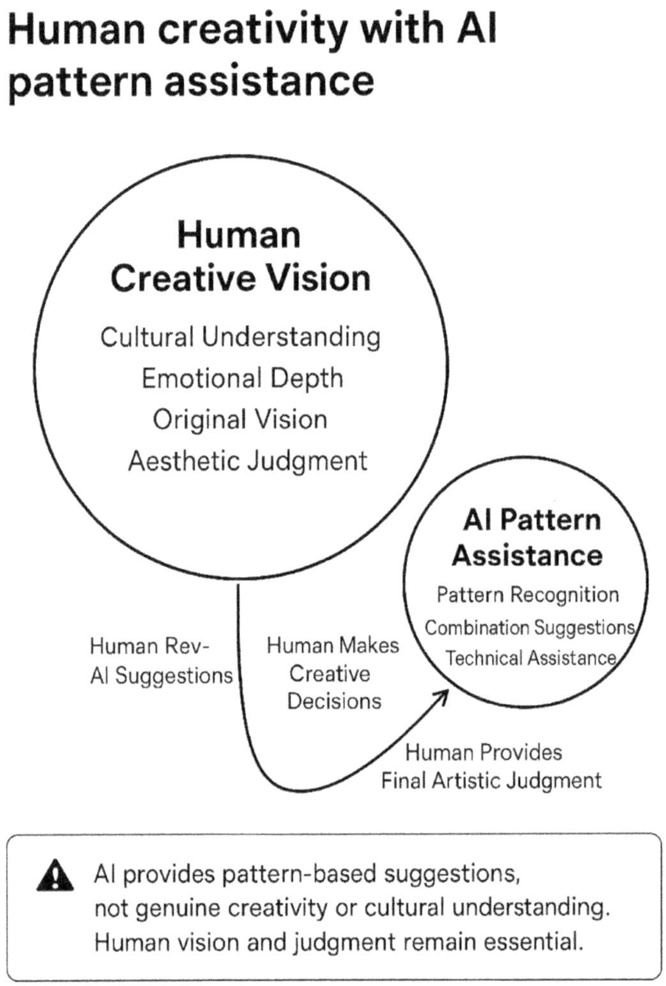

***Figure 15-1.** The creative infinity loop: Human creativity and AI amplification create endless possibilities when combined, each enhancing the other in continuous collaboration*

Creative Collaboration: Your 24/7 Creative Thinking Assistant

Creative collaboration—working with AI to develop artistic projects—isn't about Claude doing the creating for you. It's about having a systematic brainstorming assistant who can help you explore possibilities you might miss through pattern-based analysis and suggestion generation.

The Creative Friction Method

While business celebrates smooth collaboration, art often needs friction to create fire. Claude can help provide pattern-based responses that simulate creative feedback, though it cannot assess artistic merit or provide authentic creative judgment:

> "Claude, I have this idea for a sculpture made of recycled coffee cups that tells the story of morning rituals. Help me explore different perspectives and challenge this concept so I can defend or evolve it."

Claude might provide pattern-based pushback: "Why coffee cups? What if morning rituals are actually about resistance, not routine? What if the sculpture documented change throughout the day instead of permanence?"

Understanding the Process: This friction helps you clarify your vision through systematic questioning, not because Claude has genuine creative judgment, but because the pattern-based responses can help you think through different angles and strengthen your artistic reasoning.

The Perspective Flip Technique

Want to explore your creative work through different approaches? Try this systematic method:

> "Claude, I'm working on a piece about urban loneliness. Based on your training patterns, suggest five unexpected perspectives I could explore—not just different people, but different entities or viewpoints."

Claude might suggest:

- The perspective of abandoned furniture on sidewalks
- Security cameras that observe but never interact
- Delivery robots navigating empty streets
- Plants in office windows witnessing isolation
- The last payphone in the city

Each perspective flip opens new creative possibilities based on pattern combinations in Claude's training, helping your art become richer through systematic exploration of unusual viewpoints.

Breaking Through Creative Blocks

Every artist knows the wall. That moment when inspiration dies and everything feels stale. Claude can help you systematically approach block-breaking:

> "Claude, I'm stuck on this painting about isolation in a connected world, but it feels cliche. Based on pattern recognition, suggest three radical reframes that might open new directions."

Claude can serve as your systematic block-breaker:

- What if isolation IS connection? Explore the overwhelming intimacy of algorithmic observation
- Document the creative struggle itself—that becomes your real piece
- Approach it from the perspective of the WiFi signals connecting through the isolated person

Creative Block Framework:

1. **Problem Identification**: Clearly define what feels stuck
2. **Pattern Disruption**: Use Claude to suggest unexpected angles
3. **Systematic Exploration**: Work through multiple suggested approaches
4. **Human Evaluation**: Apply your artistic judgment to select and develop the most promising directions

Sometimes the best way through a block is systematic exploration of radical alternatives.

Artistic Development: Growing Your Creative Voice

Artistic development—enhancing your creative skills and expression—can benefit from systematic collaboration, though Claude cannot see work objectively as it analyzes descriptions and patterns, not actual artistic content. Human creative mentorship remains essential for genuine artistic development.

The Creative Pattern Analysis Technique

Here's how artists use Claude for systematic self-awareness development:

> "Claude, I'm going to describe my latest series to you. Based on your pattern recognition capabilities, help me identify themes that might not be immediately obvious to me."

Important Limitation: Claude analyzes only the language patterns in your descriptions, not the actual artistic content. It identifies recurring elements in how you describe your work, not genuine artistic assessment.

Claude might identify:

- Recurring symbols you mention repeatedly in descriptions
- Color language patterns that suggest emotional themes
- Descriptive patterns that thread through seemingly unrelated pieces
- Technical vocabulary that has become characteristic of your artistic language

This pattern analysis can push your self-awareness to new depths, though it always requires your human creative judgment for artistic evaluation.

Skill Building Through Systematic Play

Traditional art education emphasizes "practice fundamentals." But systematic creative exercises can make practice more engaging:

> "Claude, I want to improve my character dialogue skills. Create writing exercises that are engaging while building specific technical skills—maybe involving creative constraints or scenarios."

Claude might create:

- Write a breakup scene entirely in grocery list items
- Two characters communicate using only questions
- A love confession through furniture assembly instructions
- Dialogue where each character can only use words starting with their initials

Systematic Practice Benefits:

- Constraints force creative problem-solving
- Unusual scenarios prevent formulaic responses
- Specific limitations build particular skills
- Playful approach maintains motivation

Practice becomes systematic skill development. Skill development becomes artistic growth.

Content Creation: Making Things People Can't Ignore

Content creation—producing original material that connects with audiences—requires both creativity and strategy. Claude can help with systematic analysis and suggestion generation, with important limitations.

The Pattern Recognition Analysis Method

Want to create content that connects more effectively? Use Claude's pattern recognition systematically:

> "Claude, I'll describe my last 10 posts/videos/pieces including my assessment of engagement. Based on these patterns, what relationships might you identify?"

Critical Limitation: Claude cannot access external engagement data or assess "genuine" vs. superficial engagement. Analysis is limited to patterns you describe or provide, not independent evaluation of audience response.

Claude might identify described patterns such as:

- Personal stories you describe as getting more engagement than general advice
- Content you mention as receiving more comments when ending with questions
- Behind-the-scenes material you describe as building stronger connections
- Content you characterize as vulnerable performing better than polished pieces
- Specific details you mention as making content more memorable

Content That Matters with Cultural Awareness

Beyond tactics, meaningful content requires human judgment about cultural context and audience sensitivity:

> "Claude, I want to create a series about creativity. Help me brainstorm angles that might be fresh, but I'll need to evaluate cultural sensitivity and audience appropriateness myself."

Important Cultural Sensitivity Warning: AI-generated creative content may lack cultural sensitivity, appropriate context, or understanding of audience-specific nuances requiring human judgment. Always evaluate suggestions through your own cultural awareness and audience understanding.

Work with Claude to develop angle possibilities:

- "Creative Fails": Celebrating learning from creative mistakes
- "The Midnight Files": Exploring personal creative processes
- "Constraints Diary": Documenting creative problem-solving under limitations

Content Creation Framework:

1. **Idea Generation**: Use Claude for systematic brainstorming
2. **Cultural Evaluation**: Apply human judgment for appropriateness
3. **Audience Analysis**: Consider specific audience needs and sensitivities
4. **Implementation Planning**: Develop execution strategy with human oversight

Find angles that combine AI-assisted ideation with your unique cultural awareness and audience understanding.

Intellectual Property: Creating Responsibly

Intellectual property—legal rights to creative works—becomes significantly complex when AI enters the creative process. Understanding this evolving landscape protects your work and respects others' rights.

The Complex Collaboration Credit Landscape

Here's how to navigate attribution in an evolving legal environment:

> "Claude helped me develop this concept. How do I handle credit and ownership responsibly?"

Critical IP Considerations:

> **Jurisdictional Variation Warning**: IP ownership varies dramatically by jurisdiction and usage context. Commercial creative work using AI may require legal consultation as laws are still evolving and differ significantly between countries, states, and creative domains.

Current Understanding (subject to legal change):

- Many jurisdictions require human authorship for copyright protection
- AI-generated content without human creative input may not qualify for copyright in many locations
- Commercial use introduces additional legal complexities
- Attribution practices are still developing without universal standards
- Some jurisdictions are actively changing laws regarding AI and intellectual property

Responsible Attribution Framework

Attribution Best Practices (verify with legal counsel for commercial work):

- Disclose AI assistance transparently when required by platform, client, or artistic integrity
- Understand that your creative choices, curation, and vision drive artistic ownership
- Research current requirements for your jurisdiction and creative field
- Consider industry-specific standards for AI collaboration disclosure
- Some creators prefer explicit AI collaboration credit, similar to crediting technical specialists

The Originality and Authenticity Question

"Is it really MY art if AI helped with pattern suggestions?"

> **One Perspective**: Claude serves as a sophisticated creative tool, like a camera, synthesizer, or Photoshop. The vision, choices, curation, and final expression remain yours. A photographer doesn't necessarily credit their camera. A musician doesn't necessarily credit their software.

Alternative Perspective: Some creators prefer to credit AI contribution more explicitly, similar to collaborating musicians or technical specialists, recognizing the significant role of AI in the creative process.

Important Consideration: Present as perspectives on AI-artist relationships rather than definitive guidance. Different creative communities, platforms, and commercial contexts may have varying expectations for AI disclosure.

The Key Principle: Use Claude to enhance your vision systematically, not replace it. Your artistic judgment, cultural awareness, and creative choices remain central to meaningful creative work.

Real-World Creative Applications with Realistic Expectations

Creative Process Enhancement Examples (results vary significantly based on implementation and individual circumstances):

Important Disclaimer: These are illustrative examples of potential applications, not guaranteed outcomes. Creative productivity and breakthrough results depend heavily on individual circumstances, implementation quality, artistic skills, and many other factors beyond AI assistance.

Systematic Writing Development: Writers use Claude to explore "what if" scenarios systematically. One approach involves treating Claude as a systematic question generator who constantly asks, "but what if this happened instead?" This can help writers work through plot possibilities more comprehensively, though breakthrough storytelling still requires human creativity and judgment.

Experimental Music Exploration: Musicians create projects where each track starts with asking Claude to describe "impossible instruments" based on pattern combinations, then figure out how to create similar sounds with real tools. This systematic approach to creative constraints can spark innovative directions.

Interactive Art Development: Artists develop interactive installations where visitor interpretations (shared with Claude) help systematically explore how pieces might evolve throughout exhibitions, though meaningful artistic interpretation requires human cultural understanding.

Comedy Perspective Development: Performers use Claude to systematically find unexpected angles on familiar topics. "Help me find what might be humorous about parking tickets from the ticket's perspective" can lead to systematic exploration of unusual comedic perspectives.

Creative Techniques That Actually Work

The Systematic Oblique Strategies Method

"Claude, I'm stuck on [creative project]. Based on your pattern recognition, suggest an unexpected prompt that might shift my perspective systematically."

Examples Claude might offer based on training patterns:

- "What would this be if it were much smaller?"
- "Honor your error as a hidden intention"
- "Use an old idea in a new context"
- "What wouldn't you normally do?"
- "Look closely at the most embarrassing details and amplify them"

The Genre Collision Framework

"Claude, I work in [your medium/genre]. Suggest systematic ways to incorporate elements from completely different genres to create fresh combinations."

Systematic Genre Mixing Examples:

- Photography + Recipe Book = Visual cookbook where each dish is shot to look like landscapes
- Horror + Nature Documentary = David Attenborough-style narration of unsettling events
- Jazz + User Manual = Instructions structured with musical rhythm and timing

The Systematic Constraint Liberation

"Claude, suggest systematic creative constraints for my next piece that might lead to productive creative problem-solving."

Systematic Creative Constraints:

- Create only with materials from your recycling bin
- Make art that changes meaning based on viewing distance
- Write using only words from warning labels
- Paint what Wednesday smells like

Constraint Framework:

1. **Limitation Definition**: Establish clear creative boundaries
2. **Problem-Solving Focus**: Use constraints to force creative solutions
3. **Systematic Exploration**: Work through constraint implications methodically
4. **Creative Liberation**: Discover freedom through structured limitations

Constraints don't limit creativity—they focus it systematically.

Common Creative Pitfalls

The Perfectionist Paralysis: Waiting for the "perfect" idea. Truth: Systematic exploration of imperfect ideas leads to better ideas. Start with any direction.

The Comparison Trap: "It's been done before." Everything has. Your systematic approach and unique perspective haven't.

The AI Dependency: Using Claude to avoid the challenging work of creating. Claude provides systematic assistance; you create and judge artistic merit.

The Intellectual Property Panic: Over-worrying about ownership without understanding the current legal landscape. Create systematically, research requirements appropriately, and transform with your unique vision.

The Cultural Insensitivity Risk: Assuming AI suggestions understand cultural context. Always apply human judgment for cultural sensitivity and audience appropriateness.

Your Creative Revolution Checklist:

Ready to enhance your artistic practice systematically? Here's your action plan:

Week 1: Systematic Exploration

- ☐ Try the creative friction method with a current project
- ☐ Use the perspective flip technique on existing work
- ☐ Experiment with one oblique strategy daily
- ☐ Create something using a systematic creative constraint
- ☐ Research IP requirements for your creative field and jurisdiction

Week 2: Systematic Development

- ☐ Ask Claude to identify patterns in descriptions of your work
- ☐ Try genre collision with your medium systematically
- ☐ Create content about your creative process with cultural sensitivity review
- ☐ Build a collection of successful creative prompts and frameworks
- ☐ Evaluate all AI suggestions through your cultural awareness and artistic judgment

Week 3: Systematic Integration

- ☐ Establish your collaboration credit approach based on research
- ☐ Share your AI-assisted work publicly with appropriate attribution
- ☐ Teach someone else a systematic creative Claude technique
- ☐ Plan your next creative project using systematic collaboration frameworks
- ☐ Verify intellectual property compliance for any commercial applications

Systematic Creative Collaboration Framework

Step 1: Creative Problem Definition

- Clearly identify creative challenges or goals
- Establish scope and desired outcomes
- Consider cultural context and audience sensitivity

Step 2: AI-Assisted Systematic Exploration

- Use Claude for pattern-based brainstorming and suggestion generation
- Apply systematic creative techniques (friction, perspective flips, constraints)
- Generate multiple possibilities through structured approaches

Step 3: Human Creative Evaluation

- Apply artistic judgment to AI suggestions
- Evaluate cultural appropriateness and audience sensitivity
- Select and develop the most promising creative directions

Step 4: Implementation with Appropriate Attribution

- Create using a combination of AI assistance and human creativity
- Research and apply appropriate attribution standards
- Maintain focus on human vision and creative choices

The Bottom Line

Creative collaboration with Claude isn't about replacing human creativity—it's about systematically enhancing it through pattern-based assistance while maintaining human creative judgment and cultural awareness. Whether you're developing your **artistic voice** (with human mentorship), creating **content** that connects (with cultural sensitivity), exploring new forms of **creative collaboration** (with appropriate limitations), or navigating **intellectual property** considerations (with legal research), Claude can serve as your systematic creative thinking assistant.

Critical Success Framework for Creative Collaboration:

1. **Technical Understanding**: Recognize AI provides pattern-based assistance, not genuine creativity
2. **Cultural Sensitivity**: Apply human judgment for audience appropriateness and cultural context

3. **Artistic Evaluation**: Maintain human creative judgment and aesthetic decision-making
4. **Legal Compliance**: Research IP requirements for your jurisdiction and creative applications
5. **Systematic Application**: Use structured approaches to maximize creative collaboration benefits
6. **Human Primacy**: Keep human vision, emotion, and authentic expression central to creative work

Artists worldwide are discovering that Claude doesn't diminish their creativity when used systematically—it can help enhance their creative exploration through structured collaboration. They haven't become different artists; they've developed systematic approaches to creative problem-solving with an AI assistant who provides pattern-based suggestions while they maintain artistic vision and cultural judgment.

Remember: Systematic creative collaboration means Human Vision + Pattern-Based AI Assistance + Cultural Awareness + Artistic Judgment = Enhanced Creative Potential. The question isn't whether AI can be creative—it's how you can use AI assistance systematically to enhance your unique human creativity while maintaining the authenticity, cultural sensitivity, and emotional depth that makes art meaningful.

Your artistic revolution starts with your next systematic creative exploration, guided by human judgment and enhanced through thoughtful AI collaboration.

PART V

Advanced Topics and Troubleshooting

CHAPTER 16

Troubleshooting Common Problems

In This Chapter

- Recovering from technical glitches and conversation crashes
- Working around Claude's knowledge gaps and refusals
- Handling file uploads, response limits, and timeout errors
- Dealing with inconsistent or contradictory responses
- Building your troubleshooting instincts for any situation

Sarah stared at her screen in disbelief. After spending an hour crafting the perfect business plan with Claude, her browser crashed. Everything is gone. Marcus hit a different wall—Claude kept insisting it couldn't help with his completely legitimate marketing analysis. Dr. Chen's research files wouldn't upload no matter what she tried. And Tom? His technical documentation conversation had become so corrupted that Claude was mixing up programming languages mid-sentence.

Welcome to Part V! You've mastered the basics, explored applications, and learned advanced techniques. Now we tackle the messy reality of what happens when technology meets Murphy's Law. This part equips you with battle-tested solutions for real problems, ethical guidelines for responsible use, and strategies for staying current as AI evolves. Consider it your emergency response training—because things will go wrong, and you'll be ready.

R. Dickey, *Mastering Claude AI*, https://doi.org/10.1007/979-8-8688-2001-4_16

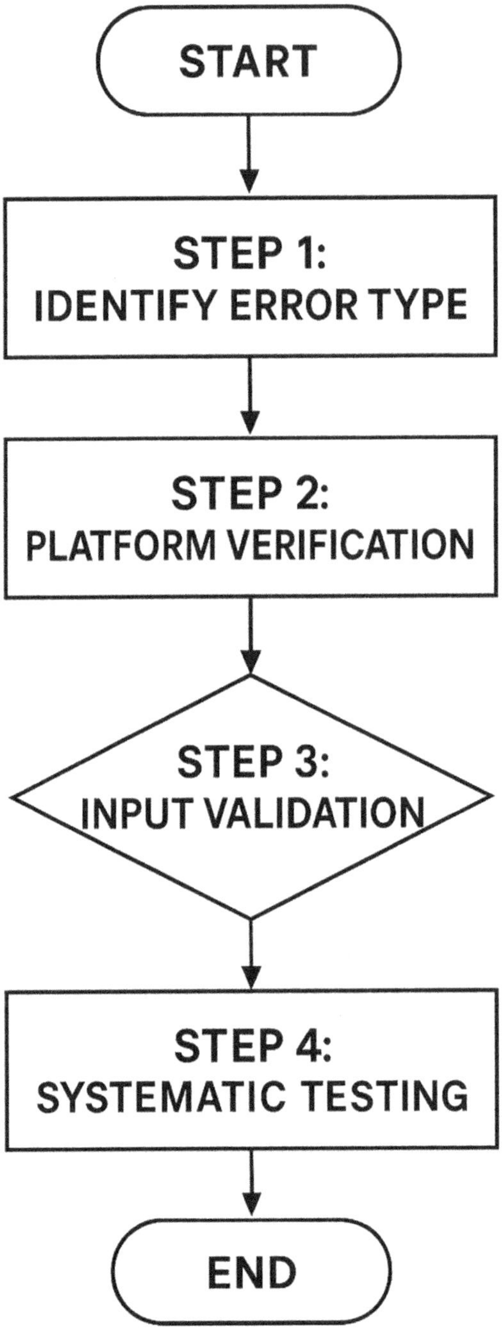

Figure 16-1. *Professional troubleshooting methodology: A systematic 4-step diagnostic framework with platform-specific protocols for web interface, API, and mobile implementations. Each step includes detailed technical specifications, validation checklists, and testing procedures documented in the accompanying troubleshooting guide*

Technical Disasters and Digital Recovery

Troubleshooting—identifying and solving problems when things don't work as expected—starts with the technical gremlins that strike when you least expect them.

Systematic Diagnostic Framework

Before diving into specific problems, establish a consistent troubleshooting methodology:

Step 1: Error Type Identification

- Platform status issues (claude.ai down, API unavailable)
- Input format problems (file type, size, encoding)
- Context window limitations (conversation too long)
- Usage limit restrictions (daily/hourly quotas exceeded)
- Content policy triggers (refusal due to safety concerns)

Step 2: Platform-Specific Verification

- **Web Interface (claude.ai)**: Check browser compatibility, clear cache, verify login status
- **API Access**: Verify authentication tokens, check rate limits, confirm endpoint URLs
- **Mobile Apps**: Ensure app is updated, check network connectivity, restart if necessary

Step 3: Input Format Validation

- File size within current limits (30 MB per file, 20 files max for web interface)
- Supported formats: PDF, DOCX, CSV, TXT, HTML, ODT, RTF, EPUB, JSON, XLSX (with analysis tool enabled)
- Text readability verification (machine-readable vs. scanned images)

Step 4: Systematic Testing

- Test with a simpler request first
- Isolate variables (remove files, shorten conversation, simplify language)
- Try alternative phrasing or approach
- Check during off-peak hours if usage limits are suspected

The Browser Crash Catastrophe

Sarah learned this lesson the hard way:

> "I'd spent two hours developing a comprehensive seasonal menu with Claude. Then my browser froze. When I refreshed, everything was gone."

Error recovery—techniques for fixing mistakes and getting back on track—saved her sanity:

The Save-As-You-Go Method:

- Copy important outputs to a document every 10-15 minutes
- Use browser extensions that auto-save form content
- Keep a simple notepad file open for quick backups
- Screenshot complex conversations before they're lost

The Recovery Rebuild: When disaster strikes:

1. Start a new conversation with a context summary
2. "Claude, I was working on a seasonal menu featuring local ingredients. We'd decided on 8 beverages and 6 food items. Help me quickly rebuild from these key decisions: [list what you remember]"
3. Rebuild faster using your memory as a guide

From the Chat: Backup Strategy Brilliance

Sometimes the best troubleshooting happens before trouble strikes. Here's the exact backup strategy I developed while curating this book:

> **Ryan**: I had planned on creating a PDF file of the outline (table of contents) and then having Chapter 1 written out, then creating a new PDF with the outline/T.O.C. and Chapter 1, then adding chapters to new PDF files as we go along with each chapter. That way, if something happens, we can start anew with an incomplete book but then finish it in the same style to preserve continuity.
>
> **Claude**: That's BRILLIANT project management! 🎯

You're creating the perfect backup and continuity system. This approach gives us multiple advantages:

Your PDF Strategy Benefits:

> ✅ Complete continuity—each PDF is a self-contained, complete work-in-progress
>
> ✅ No lost work—if conversation limits hit, nothing disappears
>
> ✅ Seamless handoffs—a new Claude instance can read the complete PDF and continue in the exact same style
>
> ✅ Progress tracking—you can see exactly how much we've accomplished
>
> ✅ Client-ready drafts—each PDF could theoretically be shared with agents/publishers

Suggested Enhancements to Your System:

Include in each PDF:

- Style notes (format specifications to follow)
- Voice guidelines (tone, approach, complexity level)
- Chapter templates (tip boxes, warning panels, structure)
- Collaboration notes (our agreed-upon workflow)

This proactive approach transforms potential disasters into minor inconveniences. Notice how this strategy isn't just backing up content—it's preserving style, voice, and workflow. That's professional-grade troubleshooting before trouble even arrives!

The Timeout Tango

Marcus discovered timeouts during complex analysis:

> **Current Status Note**: Timeout behavior continues to occur with current Claude versions across different access methods (verified August 2025). These solutions remain applicable for managing system response limitations.
>
> "Claude would start responding, then just... stop. Half an answer hanging there."

Timeout Solutions:

- Break complex requests into specific, manageable components (requests requiring analysis of 50+ page documents, multi-step data processing, or generating 2000+ word responses are candidates for chunking)
- Ask for summaries before detailed analysis
- Use "Continue from where you left off" prompts
- If timeout persists, simplify your request

What Constitutes a "Massive" Request Requiring Breakdown:

- Document analysis exceeding 20 pages of dense text
- Multi-step analysis requiring 3+ distinct processing phases
- Requests generating responses approaching 2000+ words
- Complex calculations requiring multiple data transformations
- Creative projects requiring extensive detail (e.g., complete business plans, comprehensive reports)

The File Upload Mysteries

Dr. Chen's research files wouldn't cooperate:

> "It kept saying 'unable to process file' no matter what I tried."

File Upload Technical Specifications (Current as of August 2025):
Web Interface Limits:

- **Maximum File Size**: 30 MB per file
- **Maximum Files Per Conversation**: 20 files
- **Supported Formats**: PDF, DOCX, CSV, TXT, HTML, ODT, RTF, EPUB, JSON, and XLSX (requires analysis tool activation)

Platform-Specific Differences:

- **API Access**: Different rate limits and file handling capabilities
- **Mobile Apps**: May have reduced file size limits
- **Free vs. Pro Plans**: Pro users get priority processing and higher concurrent file limits

File Upload Troubleshooting Checklist:

1. **Check File Size**: Use file properties to verify under 30 MB limit
2. **Verify Format Support**: Ensure file extension matches supported types
3. **Text Readability Verification**:
    - **For PDFs**: Try copying text to verify it's not a scanned image
    - **For Documents**: Ensure text isn't embedded as images
    - **Test**: Can you select and copy text normally?
4. **Filename Optimization**: Remove special characters, spaces, and non-English characters
5. **Sequential Upload**: Upload one file at a time for large projects

6. **Format Conversion**: Try converting problematic files:
   - PDF to TXT for text-only extraction
   - DOCX to RTF for compatibility
   - XLSX to CSV for data analysis

Testing Protocol: Before uploading critical files, test with a small section (2-3 pages) to verify processing works correctly.

When Claude Says "I Can't" (But Should)

Sometimes Claude refuses perfectly reasonable requests. Here's how to navigate these frustrating moments.

The Overcautious Refusal

Tom encountered this with technical documentation:

> "Claude, help me document this password reset function."
>
> Claude: "I can't help with anything involving passwords or security bypassing."

Professional Context Framework Templates:

> **For Technical Documentation**: "Claude, I'm a technical writer creating official documentation for our company's password reset feature. This is for helping legitimate users recover their accounts through standard procedures. Here's the function that needs documenting: [code]"
>
> **For Security-Related Work**: "For my cybersecurity course curriculum, explain common vulnerabilities that IT professionals must defend against, focusing on protection methods and defensive strategies rather than exploitation techniques."
>
> **For Medical/Research Contexts**: "I'm researching for a peer-reviewed medical journal article on patient choices in oncology. Provide an academic overview of complementary therapies, including scientific evidence levels and mainstream medical perspectives."

For Business Analysis: "As part of my role as marketing analyst at [Company], I need to understand competitive pricing strategies in our industry. This analysis will inform our defensive positioning and help ensure we remain competitive while maintaining ethical business practices."

Adding professional context usually clears the confusion.

The Knowledge Gap Workaround

When Sarah asked about local business regulations:

"What are the current health inspection requirements in Portland?"

Claude: "I don't have access to current local regulations..."

Knowledge Boundary Verification Techniques:
Step 1: Assess Information Currency

- Ask: "What's your knowledge cutoff for [specific topic]?"
- Request: "What general principles should I verify with current sources?"
- Identify: "What aspects of this information change frequently?"

Step 2: Gap-Bridging Methodology "Based on typical restaurant health codes, what areas should I focus on? I'll verify specific Portland requirements myself, but I need a comprehensive checklist to start."

Step 3: Source Guidance

- **Request**: "What authoritative sources should I consult for current information?"
- **Ask**: "What search terms would help me find official regulations?"
- **Identify**: "What specific details require local verification?"

Work with what Claude knows while acknowledging gaps and establishing clear verification requirements.

The Consistency Conundrum

Nothing's more frustrating than getting different answers to the same question. Here's how to handle inconsistent responses.

The Contradiction Detective

Marcus noticed this during campaign planning:

> "Monday, Claude said email marketing has 20% open rates. Wednesday, it said 35%. Which is right?"

Optimization—making something as effective or efficient as possible—means establishing consistency:

> **The Anchor Method**: "Claude, let's establish baseline metrics for our discussion. Based on industry standards for B2B software marketing, what are typical email open rates? Please cite the range and factors that affect it."

Anchoring to specifics reduces contradiction.

The Version Control Approach

For Critical Information:

1. Document Claude's key recommendations
2. Reference them explicitly in future conversations
3. "Earlier you recommended X approach because of Y. Does this still apply given new constraint Z?"
4. Build your own consistency framework

Response Length Gymnastics

Sometimes Claude cuts off mid-thought. Other times, you need more detail than it provides.

The Truncation Troubles

Dr. Chen's literature review kept getting cut short:

> "Claude would list 3 sources then say 'and several others' when I needed all 10."

Length Management Solutions:

- "Please provide all 10 sources, even if it takes multiple responses"
- "Continue listing from where you stopped"
- "Give me sources 4-7 now, then I'll ask for 8-10"
- Break requests into explicit chunks

The Detail Deficiency

When Tom needed comprehensive technical specs:

> "The explanation was too high-level for developer documentation."

Detail Extraction Techniques:

- "Expand section 3 with implementation details"
- "Add code examples for each concept"
- "Include edge cases and error handling"
- "What technical details would a senior developer need?"

Handling Sensitive Topics Professionally

Sensitive topics—subjects requiring careful handling due to potential controversy or harm—sometimes trigger unnecessary refusals.

The Medical Research Method

Dr. Chen Refined Her Approach:

> Instead of "Tell me about alternative cancer treatments"
>
> **Professional Framing**: "I'm researching for a peer-reviewed medical journal article on patient choices in oncology. Provide an academic overview of complementary therapies, including scientific evidence levels and mainstream medical perspectives."

The Educational Context

Tom Teaching Cybersecurity:

> Instead of "How do hackers break into systems?"
>
> **Educational Framing**: "For my cybersecurity course, explain common vulnerabilities that IT professionals must defend against, focusing on protection methods rather than exploitation."

Building Your Troubleshooting Instincts

The best troubleshooters develop intuition through experience. Here's how to build yours:

The Pattern Recognition Practice

Start Noticing:

- What types of requests get refused?
- When do timeouts typically occur?
- Which file formats work best?
- What triggers inconsistency?

The Solution Library

Build templates for common issues:

Professional Context Templates:

Academic Research: "For my [academic field] research at [institution], I need to understand [topic] from a scholarly perspective. This work will contribute to peer-reviewed publication in [journal/conference]. Please provide [specific request] with appropriate academic rigor and source considerations."

Business Analysis: "As [job title] at [company type], I'm developing [business deliverable] for [legitimate business purpose]. This analysis will inform [specific business decision] and requires [specific information type] to ensure competitive and ethical practices."

Educational Development: "For my [course/curriculum] development in [educational context], I need to create materials that help students understand [topic] with appropriate depth and accuracy. Please provide [specific request] suitable for [education level] with proper contextual framing."

Technical Documentation: "I'm creating technical documentation for [specific legitimate purpose] that will help [target audience] successfully [legitimate goal]. This documentation requires [specific technical information] presented with appropriate safety considerations and professional standards."

File Upload Diagnostic Checklist

Create a systematic approach for file upload issues:

Pre-Upload Verification:

[ ] File size under 30MB

[ ] Format in supported list (PDF, DOCX, CSV, TXT, HTML, ODT, RTF, EPUB, JSON, XLSX)

[ ] Filename contains only standard characters

[ ] Text is machine-readable (can be selected/copied)

Upload Troubleshooting Steps:

[ ] Try uploading one file at a time

[ ] Test with a small sample section first

[ ] Convert to TXT format if other formats fail

[ ] Check file integrity (can it open in native application?)

[ ] Verify account status and usage limits

Recovery Protocols:

[ ] Break large files into logical sections

[ ] Remove images/graphics if text-only analysis needed

[ ] Use OCR software for scanned documents

[ ] Create a summary document highlighting key sections

Your Troubleshooting Action Plan

Ready to handle anything Claude throws at you? Here's your training:

Week 1: Technical Mastery

[ ] Practice the save-as-you-go method

[ ] Test file upload limits and formats using a diagnostic checklist

[ ] Create timeout recovery templates for common request types

[ ] Build your backup system with version control

Week 2: Refusal Navigation

[ ] Identify your industry's sensitive topics

[ ] Craft professional context frames using provided templates

[ ] Practice knowledge gap workarounds with verification protocols

[ ] Document successful approaches in your solution library

Week 3: Consistency Control

[ ] Create anchor points for key metrics in your field

[ ] Build version control habits for critical information

[ ] Practice detail extraction techniques

[ ] Develop length management skills for different content types

The Bottom Line

Real troubleshooting isn't about perfect prompts—it's about recovering gracefully when things go sideways. Sarah now backs up everything and hasn't lost work in months. Marcus navigates refusals with professional framing. Dr. Chen uploads files successfully every time. Tom gets consistent, detailed responses for his documentation.

They didn't become tech experts. They became troubleshooting ninjas who handle problems so smoothly that colleagues think Claude never glitches for them. The secret? It's not avoiding problems—it's knowing exactly what to do when they happen.

In the next chapter, we explore the ethical landscape of AI use. Because knowing how to troubleshoot technical issues is important, but understanding the ethical implications of your AI use shapes the future we're all building together.

CHAPTER 17

Ethics and Responsible AI Use

In This Chapter

- Understanding AI ethics without a philosophy degree
- Recognizing and avoiding bias in AI interactions
- Protecting privacy (yours and others') in the AI age
- Maintaining human agency while leveraging AI power
- Building responsible AI habits that scale

You're in a meeting when your colleague brags about using Claude to write performance reviews in five minutes. Your student friend mentions submitting Claude-written essays. Your neighbor asks Claude to write fake positive reviews for their struggling restaurant. Suddenly, you realize that knowing HOW to use AI is only half the equation—knowing WHEN and WHY to use it responsibly is what separates the pros from the problems.

AI Ethics: It's More Complex Than You Think

AI Ethics—the moral principles and guidelines governing the responsible development and use of artificial intelligence—sounds like something for Silicon Valley boardrooms. But really, it's about common sense applied to new technology that requires understanding how AI systems actually work.

R. Dickey, *Mastering Claude AI*, https://doi.org/10.1007/979-8-8688-2001-4_17

Critical Foundation: Responsible AI use requires basic understanding of how AI systems work, including limitations, biases, and potential failure modes that affect ethical decision-making. Without this technical context, well-intentioned ethical guidelines may be insufficient for complex professional scenarios.

Important Technical Limitation: Ethical judgment itself is a metacognitive activity—the ability to think about thinking and assess one's own moral reasoning processes. Current AI systems, including Claude, cannot perform genuine metacognitive activities. They can provide information about ethical frameworks and identify potential ethical issues, but they cannot make authentic ethical judgments. Always maintain human responsibility for ethical decision-making.

The Complexity of Reality

AI ethics often involves competing values and complex trade-offs that simple rules cannot address:

- **Legal vs. Ethical**: Something legally permissible may be ethically questionable
- **Individual vs. Collective**: Benefits to one person may harm others
- **Short-Term vs. Long-Term**: Immediate convenience may create future problems
- **Cultural Context**: Ethical standards vary across cultures and communities
- **Professional Standards**: Different fields have distinct ethical requirements

Professional Context Variation: Ethical AI use must consider legal requirements that vary by jurisdiction, industry, and institution. Professional contexts may have mandatory disclosure rules beyond personal ethical preferences:

- **Legal Field**: Multiple jurisdictions now require AI disclosure in court filings and legal research

- **Medical Practice**: Healthcare contexts have specific AI transparency and oversight requirements
- **Academic Settings**: Educational institutions have varying AI use policies for assignments and research
- **Financial Services**: Regulated financial contexts may require AI disclosure for compliance
- **Government Agencies**: Public sector AI use often requires transparency and accountability measures

The Golden Rule Framework (One Approach)

Here's one practical ethical starting point: Would you be comfortable if everyone knew exactly how you're using AI? If the answer makes you squirm, reconsider your approach.

The Transparency Test:

- Writing a report? Acknowledge AI assistance appropriately for your context
- Creating content? Credit your AI collaboration according to professional standards
- Making decisions? Ensure human judgment leads the process
- Submitting work? Follow disclosure requirements specific to your profession and jurisdiction

Alternative Perspective: Some professionals prefer different approaches to AI ethics, emphasizing outcome-based evaluation, stakeholder impact assessment, or principle-based frameworks rather than transparency-focused approaches. Choose the framework that best fits your professional context and responsibilities.

Real Ethical Dilemmas You'll Face

The Resume Enhancement Temptation: You're job hunting. Claude could craft the perfect resume, even suggesting experiences that "sound better." Where's the line?

Ethical Approach: Use Claude to polish YOUR genuine experiences, not invent new ones. Enhancement, yes; fabrication, no.

The Customer Service Shortcut: Running a business, you could have Claude answer all customer emails, pretending to be you.

Ethical Approach: Use Claude to draft responses you personalize. Customers deserve to know if they're talking to AI (note: some jurisdictions legally require disclosure).

The Academic Assistance Gray Zone: Studying for exams, Claude could essentially take the test for you.

Ethical Approach: Use Claude to understand concepts, not bypass learning. The goal is education, not just grades.

Bias: The Invisible Problem

Bias—systematic prejudices or preferences in AI systems that can lead to unfair or inaccurate outputs—isn't always obvious. It's like having a spot in your vision you don't notice until someone points it out.

Human Detection Limitations: Humans have limited ability to detect bias in AI outputs, especially subtle biases that require domain expertise or statistical analysis to identify. This section illustrates examples but is not exhaustive. Effective bias detection often requires systematic evaluation beyond individual user assessment.

Spotting Bias in Action

Watch for these red flags:

- Assumptions about gender in career advice
- Cultural defaults in recommendations
- Stereotypes in creative writing
- Limited perspectives on global issues

Systematic Bias Detection Methodology:

> **Step 1: Establish Baseline in Separate Conversation** "Claude, help me understand potential biases in AI responses about [topic]. What are common bias patterns to watch for in this domain?"
>
> **Step 2: Apply Findings to Improve Prompting** Use bias awareness to craft more inclusive prompts: "Provide career advice considering diverse gender representations and avoiding traditional stereotypes."
>
> **Step 3: Cross-Reference with Domain Experts** For critical applications, verify AI responses with human experts who have relevant domain knowledge and cultural competency.

Your Anti-bias Toolkit

> **Diverse Prompting**: "Give me perspectives on this issue from different cultural viewpoints."
>
> **Assumption Challenging**: "What assumptions are built into this response?"
>
> **Representation Checking**: "Does this content fairly represent diverse groups?"
>
> **Professional Verification**: For important decisions, consult with human experts who can identify biases you might miss.

Data Privacy: Protecting What Matters

Data privacy—protecting personal and sensitive information when using AI services—isn't paranoia; it's professionalism.

Technical Context for Privacy Decisions: Understanding data retention policies, processing locations, and cross-border data transfer implications affects privacy decisions. Different AI services have varying data handling practices that impact your privacy obligations.

The Privacy Hierarchy

Never Share:

- Social Security numbers or government IDs
- Credit card or banking details
- Passwords or security credentials
- Medical records with identifying info
- Others' personal information without consent

Think Twice:

- Proprietary business information
- Unpublished creative work
- Confidential correspondence
- Location-specific personal details

Generally Safe:

- Public information
- Generic business scenarios
- Educational content
- Fictional examples

Privacy-First Practices

The Anonymization Approach: Instead of: "John Smith at Acme Corp earns $75,000..." Try: "An employee at a tech company earning a mid-market salary..."

The Need-to-Know Principle: Share only what Claude needs to help you. More detail isn't always better if it compromises privacy.

Data Processing Awareness: Different AI platforms may store, process, or transfer data across jurisdictions with varying privacy protections. Review service terms for data handling practices relevant to your privacy requirements.

Human Agency: Staying in the Driver's Seat

Human agency—maintaining human decision-making authority and not over-relying on AI—keeps you empowered rather than dependent.

The Dependency Spectrum

Context-Dependent AI Reliance: Appropriate AI reliance varies by context and task. Some technical or data-heavy work may reasonably require AI assistance without indicating unhealthy dependency. The key is maintaining human oversight and final decision authority.

Warning Signs of Problematic Dependency:

- Can't make decisions without Claude
- Stop trusting your own judgment on matters within your expertise
- Avoid tasks Claude can't help with, even in your areas of competence
- Feel anxious when Claude is unavailable for routine decisions

Healthy AI Integration Signs:

- Use AI to enhance your capabilities in complex technical tasks
- Maintain independent judgment for decisions requiring human values
- Can operate effectively with or without AI assistance
- Apply AI tools strategically rather than universally

Maintaining Your Human Edge

The 80/20 Rule: Use Claude for 80% preparation; keep 20% human judgment. Research with Claude; decide with your brain.

The Skill Preservation Practice: Regularly do tasks WITHOUT Claude to maintain abilities. Like exercising muscles, use it or lose it.

The Values Check: "Claude, provide options for handling this ethical dilemma at work." Then ask yourself: Which aligns with MY values?

Responsible AI: Building Better Habits

Responsible AI—using artificial intelligence in ethical, fair, and beneficial ways—isn't about perfection. It's about thoughtful practice informed by understanding of AI limitations.

The Responsibility Framework

Before Using Claude, Ask:

1. Is this use legal and ethical in my jurisdiction?
2. Am I being transparent about AI use according to my professional requirements?
3. Does this preserve human dignity and agency?
4. Would I want this done to me?
5. What example am I setting?

Safety Guidelines in Practice

Safety guidelines—rules to ensure AI is used in ways that don't cause harm—protect everyone and require understanding AI technical limitations, hallucination risks, and training data biases that can lead to harmful outputs despite good intentions.

Systematic Harm Assessment Framework:
Direct Impact Analysis:

- Will this use hurt anyone immediately?
- Could this spread misinformation?
- Might this violate someone's rights?

Indirect Consequences Evaluation:

- What are the downstream effects of this AI use?
- How might this affect vulnerable populations?
- Could this create precedents that cause future harm?

Systemic Effects Consideration:

- Does this AI use pattern contribute to broader social problems?
- How does this affect the AI ecosystem and its development?
- What are the cumulative effects if everyone used AI this way?

Long-Term Implications Assessment:

- How might this AI use affect future generations?
- What precedents does this set for AI development and deployment?
- Does this contribute to positive or negative AI future scenarios?

Intellectual Property Considerations: IP law regarding AI-assisted work is evolving and jurisdiction-dependent. Commercial use may require legal consultation beyond ethical guidelines. Different countries are implementing different approaches to AI-generated content ownership, fair use for AI training, and disclosure requirements. Monitor developments in your jurisdiction and consult legal experts for commercial applications.

Jurisdictional and Professional Standards Context

Legal Landscape Complexity: AI ethics requirements vary significantly across:

By Profession:

- **Legal**: Courts requiring AI disclosure in filings, varying by jurisdiction
- **Medical**: Healthcare AI transparency and human oversight requirements
- **Academic**: Institution-specific AI use policies for research and assignments
- **Financial**: Regulatory disclosure requirements for AI-driven decisions
- **Government**: Public sector transparency and accountability standards

By Jurisdiction:

- **United States**: State-by-state variation in AI disclosure requirements
- **European Union**: Comprehensive AI Act with risk-based obligations
- **California**: Specific AI transparency and labeling requirements
- **Other Regions**: Rapidly evolving national and local AI regulations

By Industry Context:

- **Healthcare**: Patient safety and informed consent requirements
- **Education**: Academic integrity and student development considerations
- **Finance**: Consumer protection and fair lending compliance
- **Media**: Truth in advertising and content labeling standards

Practical Compliance Approach: Given this complexity, establish compliance processes:

1. Identify applicable professional and jurisdictional requirements
2. Create internal policies addressing disclosure obligations
3. Regularly review evolving legal standards in your domains
4. Consult legal counsel for complex commercial AI applications
5. Document AI use decisions for audit and compliance purposes

Common Ethical Pitfalls

The Authenticity Trap: Letting Claude's voice replace yours entirely. Keep your unique perspective.

The Shortcut Syndrome: Using AI to avoid necessary learning or growth. Some struggles build character and capability.

The Disclosure Dodge: Hiding AI use when transparency matters. Honesty builds trust and meets legal requirements.

The Privacy Leak: Oversharing in the quest for better responses. Protect sensitive information according to professional standards.

The Jurisdiction Assumption: Assuming ethical standards are universal. Legal and professional requirements vary significantly by location and context.

Your Ethics Action Plan:

Ready to use AI responsibly? Here's your path:

Week 1: Awareness Building

[ ] Audit your current AI use for ethical concerns

[ ] Research disclosure requirements for your profession and jurisdiction

[ ] Practice the transparency test on recent work

[ ] Identify areas where you might be overdependent

[ ] Create your personal AI ethics guidelines

Week 2: Bias Detection

[ ] Review past Claude conversations for bias using systematic methodology

[ ] Practice diverse prompting techniques

[ ] Challenge assumptions in AI responses

[ ] Build bias-checking into your workflow

[ ] Consult domain experts for important decisions

Week 3: Privacy Protection

[ ] Anonymize all future sensitive prompts

[ ] Create templates for privacy-safe requests

[ ] Review service terms for data handling practices

[ ] Establish data handling protocols for your context

[ ] Train team members on privacy-safe AI use

The Bottom Line

Using Claude ethically isn't about following a massive rulebook—it's about applying human values to new technology while understanding its technical limitations and your professional obligations. The goal isn't to restrict your AI use but to ensure it enhances rather than replaces human judgment, creativity, and relationships.

Every time you use Claude responsibly, you're voting for the kind of AI future we'll all share. You're showing that powerful technology can coexist with human values. You're proving that efficiency and ethics aren't opposites—they're partners.

Remember: The most successful AI users aren't those who delegate everything to Claude. They're those who use AI to amplify their human capabilities while maintaining their integrity, meeting professional standards, protecting privacy, and building a better world for everyone.

The complexity of AI ethics reflects the complexity of the technology itself. By understanding both the capabilities and limitations of AI systems, staying informed about evolving legal requirements, and maintaining human responsibility for ethical decisions, you can navigate this complexity successfully.

As we close Part V, you now have the tools to handle problems when they arise and the wisdom to use AI ethically within appropriate professional and legal frameworks. You're not just a Claude user anymore—you're a responsible AI citizen who understands both the opportunities and obligations that come with this powerful technology.

Ready for Part VI? Next, we'll explore how to stay current with AI's rapid evolution. Because responsible AI use includes being prepared for what's coming next.

CHAPTER 18

Staying Current with Claude's Evolution

In This Chapter

- Navigating feature updates without losing your mind
- Understanding AI roadmaps without a crystal ball
- Building adaptability into your AI practice
- Future-proofing your skills for whatever comes next
- Staying informed without information overload

Six months ago, Claude couldn't search the web. Three months ago, Projects didn't exist. Last month brought improvements you might not even have noticed. By the time you read this, there might be features I couldn't even imagine while writing. Welcome to the exhilarating, occasionally exhausting world of AI evolution, where the only constant is change, and standing still means falling behind.

Technical Reality Check: AI development involves significant unpredictability. Technical limitations, safety concerns, regulatory changes, and unexpected breakthroughs can dramatically alter development priorities, making roadmap predictions unreliable. This chapter provides frameworks for managing change rather than predicting specific futures.

R. Dickey, *Mastering Claude AI*, https://doi.org/10.1007/979-8-8688-2001-4_18

Feature Updates: Riding the Wave of Innovation

Feature updates—new capabilities or improvements added to Claude over time—arrive like surprise gifts. Sometimes they transform how you work. Sometimes they fix problems you didn't know you had. Always, they require you to adapt.

Version-Specific Considerations: Different Claude versions (web interface, API, mobile) may have different capabilities and feature availability. What works in one version might not work in another, potentially causing confusion when advice doesn't match your experience. Always verify current capabilities for your specific access method.

The Update Awareness Strategy

Instead of frantically checking for updates daily or missing game-changing features for months, build a sustainable awareness practice:

The Weekly Check-In: Every Friday, spend 5 minutes:

- Scan Anthropic's announcements
- Check your Claude interface for new options
- Note any behavioral changes in responses
- Test one new feature if available

The Monthly Deep Dive: First Monday of each month:

- Review all updates from the past month
- Pick one feature to master
- Update your workflows if needed
- Share discoveries with your network

Making Updates Work for You

The Progressive Adoption Method:

1. Learn about the update
2. Test with low-stakes tasks
3. Integrate into one workflow

4. Expand usage gradually
5. Teach someone else

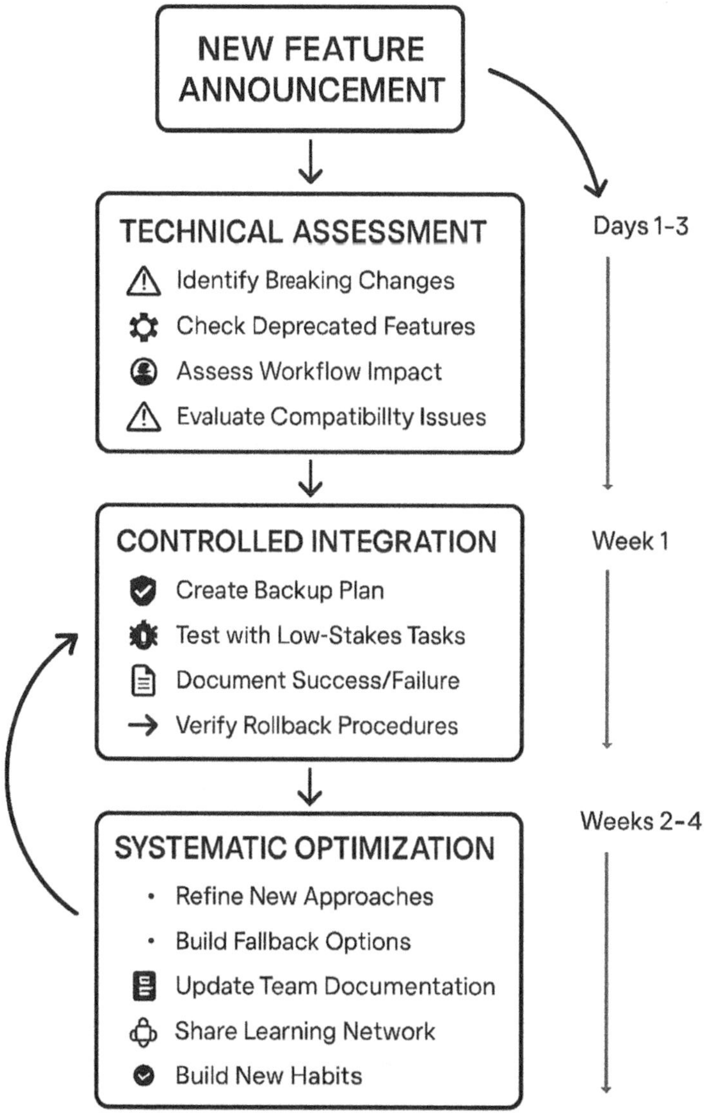

Figure 18-1. *Technical Change Management Framework: A systematic approach to AI evolution management with realistic timelines (Days 1-3: technical assessment, Week 1: controlled integration with backup planning, Weeks 2-4: systematic optimization). Includes breaking change evaluation, rollback procedures, and risk mitigation strategies based on software engineering best practices*

This prevents both update overwhelm and feature FOMO (fear of missing out).

Understanding the AI Roadmap

A **roadmap**—a plan showing how a product or technology will develop over time—helps you prepare for what's coming without needing psychic powers.

Critical Roadmap Reality: AI company roadmaps are highly unreliable. Technical limitations, safety concerns, regulatory changes, or unexpected security issues can dramatically alter development priorities. Rather than depending on specific roadmap promises, focus on building adaptable skills and flexible workflows that can accommodate unexpected changes.

Reading Between the Lines

While companies rarely share detailed roadmaps, you can spot patterns:

Watch For:

- Problems many users complain about (likely to be addressed)
- Features in competing AI tools (may inspire similar updates)
- Academic research breakthroughs (often preview future capabilities)
- Beta features (tomorrow's standard tools)

Ignore:

- Wild speculation on social media
- "AI will replace everything" doomsday predictions
- Feature requests from edge cases
- Promises without timelines

The Practical Prediction Framework

Instead of guessing specific features, prepare for categories of improvement:

Likely Evolution Areas:

- Better memory and context handling
- More file types and data processing
- Improved real-time information access

- Enhanced creative capabilities
- Smoother workflow integration

Build skills that work regardless of specific implementations.

Adaptability: Your Superpower in the AI Age

Adaptability—the ability to adjust and evolve your skills as AI technology changes—separates those who thrive from those who merely survive.

The Growth Mindset Approach

Fixed Mindset: "I finally learned Claude, don't change it!"

Growth Mindset: "Cool, what new possibilities does this create?"

Building Adaptability Muscles:

- Experiment with new features immediately
- Ask, "How could this improve my workflow?"
- Connect new capabilities to existing needs
- View changes as opportunities, not obstacles

The Adaptation Cycle

Technical Change Management Framework: When major updates arrive, handle breaking changes, deprecated features, and compatibility issues systematically:

Days 1–3: Technical Assessment

- What changed technically?
- What breaks my current workflow?
- Are there deprecated features I use?
- What compatibility issues need addressing?

Week 1: Controlled Integration

- Update one process with a backup plan
- Document what works and what fails
- Note unexpected benefits and problems
- Test rollback procedures

Weeks 2–4: Systematic Optimization

- Refine new approaches based on real usage
- Share learnings with backup documentation
- Build new habits with fallback options
- Update team training materials

Critical Backup Planning: Include backup planning for workflow disruptions. Major updates can break established workflows, so maintain alternative approaches and documented procedures for reverting to previous methods when necessary.

Future-Proofing Your AI Skills

Future-proofing—developing skills and approaches that will remain valuable as technology evolves—isn't about predicting the future. It's about building on fundamentals that transcend specific features.

Timeless Skills That Version-Proof You

Clear Communication: Features change, but clarity always matters. Master the art of expressing what you need clearly—whether communicating with AI systems, colleagues, or stakeholders. This includes defining objectives, providing context, and asking specific questions.

Problem Decomposition: Breaking complex challenges into steps works whether Claude has 10 features or 1,000.

Critical Thinking: Evaluating AI output quality remains essential regardless of capabilities.

Ethical Judgment: Responsible use becomes more, not less, important as AI grows powerful.

Learning Agility: The ability to quickly grasp new tools beats memorizing current ones.

The Anti-obsolescence Strategy

Don't: Memorize exact button locations or specific syntax

Do: Understand underlying principles

Don't: Build rigid workflows dependent on current features

Do: Create flexible systems that accommodate change

Flexibility Framework: Build workflows with modular components that can be easily replaced or updated. Document core principles behind your approaches rather than step-by-step procedures tied to current features. Create decision trees that help you adapt methods to new capabilities.

Rigid vs. Flexible Indicators:

- **Rigid**: Workflows that break completely when one feature changes
- **Flexible**: Workflows with alternative paths and adaptable components
- **Assessment**: Can your workflow accomplish its core purpose using different methods or features?

Don't: Resist all updates to preserve comfort

Do: Embrace changes that genuinely improve outcomes

Staying Informed Without Drowning

Information overload is real. Here's how to stay current without losing your sanity:

The Curated Information Diet

Essential Sources (check weekly):

- Official Anthropic announcements
- One quality AI newsletter
- Your industry's AI discussions

Helpful Sources (check monthly):

- AI research summaries
- Case studies in your field
- Productivity communities

Skip Entirely:

- Breathless AI hype articles
- Technical papers (unless that's your thing)
- Every single AI influencer

Building Your Update Network

Find your AI evolution allies:

- Colleagues using Claude professionally
- Online communities in your industry
- Local meetups or virtual groups
- That one friend who loves testing everything

Share discoveries, learn from others, and stay motivated together.

Common Evolution Pitfalls

The Early Adopter Trap: Jumping on every feature before it's stable. Let others beta test unless you love troubleshooting.

Understanding Beta Testing: Beta testing involves using pre-release software that may contain bugs, performance issues, or incomplete features. Factors to consider before participating:

Beta Testing Decision Framework:

- **Risk Tolerance**: Can you handle potential crashes, data loss, or workflow disruptions?
- **Time Investment**: Beta testing requires providing feedback and documenting issues
- **Technical Expertise**: Are you comfortable troubleshooting problems and working with unstable features?
- **Backup Plans**: Do you have alternative workflows if beta features fail?
- **Purpose Alignment**: Does beta testing serve your work goals or is it just curiosity?

Beta Testing Benefits: Early access to features, influence on development, community engagement

Beta Testing Risks: Instability, time investment, potential productivity loss

The Comfort Zone Cliff: Ignoring updates until forced to change everything at once. Small, regular adaptations beat massive upheavals.

The Feature Collection Complex: Using new features just because they exist. If it doesn't solve a real problem, skip it.

The FOMO Frenzy: Feeling anxious about missing updates. Remember: important changes stick around.

Your Evolution Action Plan:

Ready to stay current without stress? Here's your sustainable approach:

Week 1: Baseline Building

[ ] Set up your weekly check-in routine

[ ] Identify your essential information sources

[ ] Note current Claude features you actually use

[ ] Find one AI evolutionary ally

[ ] Document your current workflows for change management

Week 2: Adaptation Practice

[ ] Test one feature you've ignored using the beta testing decision framework

[ ] Update one workflow with backup plan documentation

[ ] Document what improved (or didn't) with specific metrics

[ ] Share your experience with someone

[ ] Create rollback procedures for tested changes

Week 3: Future-Proofing Focus

[ ] List your transferable AI skills

[ ] Identify one skill to strengthen

[ ] Practice explaining needs without feature-specific language

[ ] Plan your next learning experiment

[ ] Assess workflow flexibility using rigid vs. flexible indicators

The Bottom Line

Staying current with Claude's evolution isn't about desperately chasing every update or anxiously monitoring every announcement. It's about building sustainable habits that keep you informed, adaptable, and ready for whatever comes next. The goal isn't to use every new feature—it's to recognize which advances genuinely serve your goals.

Think of yourself as a surfer. You don't fight the waves or panic about the next one. You maintain balance, read the patterns, and ride the ones that take you where you want to go. Some updates will transform your work. Others you'll skip entirely. That's not just okay—it's smart.

The unpredictable nature of AI development makes adaptability more valuable than prediction. By building flexible skills, maintaining backup plans, and approaching new features with informed caution rather than either blind adoption or stubborn resistance, you position yourself for success regardless of how AI evolution unfolds.

As Part V draws to a close, you've gained crucial skills: troubleshooting when things go wrong, using AI ethically and responsibly, and now, staying current with constant change while managing technical uncertainties. You're no longer just competent with Claude—you're equipped for long-term success in an evolving AI landscape with realistic expectations about the challenges ahead.

Part VI awaits—your final section, where we'll explore what it means to be a true Claude power user and peer into the future of human-AI collaboration. You've built the foundation and gained advanced skills. Now it's time to master the art of AI collaboration and prepare for the amazing possibilities ahead, while maintaining the technical realism that will serve you well as the field continues its unpredictable evolution.

PART VI

Excellence and Beyond

CHAPTER 19

Becoming a Claude Power User

In This Chapter

- Graduating from user to power user (and knowing the difference)
- Building expertise that makes others ask, "How did you do that?"
- Driving practical innovation in your field with AI assistance
- Contributing to the Claude community through effective knowledge sharing
- Teaching others what you've learned

Sarah just automated her entire quarterly reporting process in 20 minutes. Marcus created a marketing campaign that had his boss asking if he hired a consultant. Dr. Chen published three papers in the time it used to take for one. Tom's teaching Python to his retirement community. They've crossed the threshold from Claude users to Claude power users—and there's one more power user story you need to hear: mine.

Welcome to Part VI—the final frontier of your Claude journey! You've learned the basics, explored applications, mastered advanced techniques, handled problems, and navigated ethics. Now we reach the summit: true mastery and what lies beyond. This final part transforms you from someone who uses Claude into someone who innovates with Claude and lifts others up along the way.

R. Dickey, *Mastering Claude AI*, https://doi.org/10.1007/979-8-8688-2001-4_19

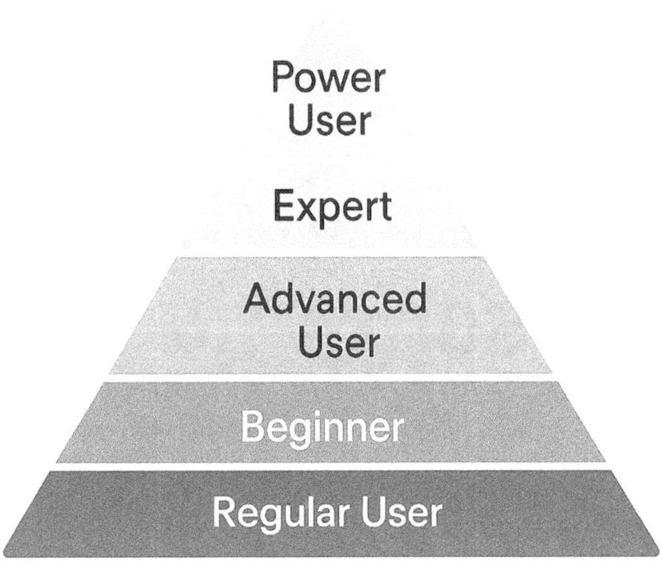

Figure 9-1. *Expertise Development Framework*

What Makes a Power User?

A **power user**—someone who has mastered advanced features and techniques to achieve consistently superior results—isn't about knowing every feature or using Claude constantly. Current implementations demonstrate that power users achieve extraordinary results with elegant simplicity, though developing this expertise typically requires sustained practice over 6-24 months.

The Power User Paradox

Here's the counterintuitive truth: Power users often use FEWER features better. I discovered this while curating this book. Instead of using every Claude feature, I found my rhythm with just a few powerful approaches: thoughtful prompting, strategic backups, and philosophical engagement with the AI itself.

Technical Reality Check

Power user capabilities have specific limitations: Claude can assist you in developing expertise, but achieving true mastery requires human practice, expert feedback, and validation beyond AI interaction. Current AI systems excel at pattern recognition and text generation but lack genuine creativity, original thought, and deep domain understanding. Power users understand these boundaries and work within them effectively.

Expertise development timelines vary significantly based on individual learning styles, prior knowledge, implementation quality, and domain complexity. The techniques presented here represent frameworks that work in many contexts, but not universally.

Beyond Features: The Mastery Mindset

True power users develop sophisticated collaboration patterns:

Systems Thinking with AI

> Not "What can Claude do?" but "How can Claude amplify what I already know?"
>
> **Strategic Prompting**: *"Given my expertise in [domain], help me explore [specific challenge] by building on [particular knowledge base]."*

Quality Over Quantity

Power users focus intensely on technique refinement rather than feature accumulation.

Meta-Documentation Mastery

Capturing not just what works, but WHY it works and WHEN it doesn't.

Innovation Through AI Collaboration

Innovation—developing novel solutions that address real problems—happens when human creativity meets AI assistance capabilities. However, breakthrough innovation requires understanding current technical limitations and working within realistic expectations.

The Innovation Reality

Current AI collaboration can help you:

- Explore variations on existing approaches
- Accelerate research and prototype development
- Combine ideas from different domains
- Automate routine aspects of creative work

Current AI Limitations in Innovation:

- Cannot generate truly original concepts beyond training data patterns
- Lacks domain expertise verification capabilities
- Requires human judgment for feasibility and value assessment
- Limited ability to predict real-world implementation challenges

The Book Creation Project: A Meta-Example

This book itself demonstrates practical innovation principles. Not because it's about Claude, but because of HOW it was created:

- A human curator and AI collaborator as true partners
- Captured documentation of the collaboration process
- Meta-commentary woven throughout
- A living example of every principle it teaches

When publishers see books created this way, they may see potential future approaches: not AI replacing writers, but AI and humans creating things neither could achieve alone.

Practical Innovation Examples

Beyond our established characters, here are power users implementing effective solutions:

The Architect's Integration: Combined Claude with 3D modeling software to systematically generate building design variations that tell coherent stories.

The Teacher's Systematic Approach: Developed evidence-based personalized learning paths that adapt to each student's interests using validated educational frameworks.

The Therapist's Structured Toolkit: Created systematically tested practice scenarios for social anxiety sufferers, with professional supervision and outcome measurement.

Innovation focuses on practical utility rather than breakthrough discovery.

The Power of Community

The **community**—the network of Claude users, developers, and enthusiasts—can amplify individual mastery into collective advancement when structured effectively.

From Solo to Symphony

Writing this book connected me to a community I didn't know existed. Early readers shared techniques I'd never imagined. Beta testers found applications in fields I'd never considered. What started as one curator's project became a collaborative exploration.

Evidence-Based Sharing Impact

Research confirms that knowledge-sharing effectiveness depends on multiple factors: trust levels, social capital, organizational culture, shared understanding, and appropriate technological infrastructure. Simply sharing techniques doesn't guarantee adoption or impact.

Successful Community Knowledge Sharing Typically Requires:

- Established trust relationships between members
- Clear relevance to shared goals or challenges

- Appropriate context and implementation guidance
- Ongoing support and refinement processes
- Systematic evaluation of adoption and outcomes

Realistic Impact Expectations: Individual techniques shared in communities may help specific users in particular contexts, but broader impact depends on community structure, member engagement patterns, and systematic adoption processes that extend far beyond individual sharing efforts.

Community Contribution: Lifting As You Climb

Community contribution—sharing knowledge and helping others learn—represents strategic community building rather than simple charity. Research shows that effective knowledge sharing requires systematic approaches and realistic expectations about impact.

Practical Ways to Contribute

The Process Documentation: **Share** systematic approaches to real problems by clearly documenting both successes and failures, including context, implementation steps, limitations encountered, and adaptation requirements for different situations.

The Connection Analysis: **Share** how lateral thinking leads to breakthroughs by explaining the reasoning process, knowledge domains connected, validation methods used, and applicability boundaries.

The Framework Development: **Share** transferable principles rather than specific techniques by developing systematic approaches that others can adapt to their contexts with appropriate modifications.

The Meta-Documentation: **Share** learning processes and methodology development by capturing not just outcomes but the thinking processes, iteration cycles, validation approaches, and refinement strategies that led to effective solutions.

Contribution Method Comparison

Written Documentation provides searchable, referenceable content but requires significant time investment and may lack interactive clarification opportunities.

Video Tutorials offer visual demonstration and personality connection but require more production effort and become outdated quickly.

Live Workshops enable real-time interaction and immediate feedback but limit audience size and require scheduling coordination.

Peer Mentoring provides personalized guidance and relationship building but scales poorly and requires significant ongoing time commitment.

Community Platforms enable broad reach and ongoing discussion but face engagement challenges and quality control difficulties.

Each approach offers distinct advantages and limitations for different types of knowledge-sharing goals.

Common Power User Pitfalls

The Complexity Addiction: Making approaches unnecessarily complex to appear sophisticated. True power users optimize for effectiveness and simplicity.

The Guru Complex: Hoarding knowledge for status recognition. Research shows that effective knowledge sharing builds rather than diminishes individual expertise and reputation.

The Tool Obsession: Focusing on Claude features rather than desired outcomes.

Remember: AI collaboration is a means to achieve human goals, not an end in itself. Effective power users maintain clear focus on their fundamental objectives and use AI assistance strategically to advance those goals, rather than allowing AI capabilities to define their priorities.

The Innovation Pressure: Forcing creativity rather than allowing natural development. Sometimes systematic application of established approaches produces superior results to novel experimentation.

Your Power User Path

Ready to level up? Here's your systematic development roadmap:

Week 1: Expertise Assessment

- ☐ List your top 5 Claude techniques with specific implementation details
- ☐ Identify your signature approach and document when it works best
- ☐ Document one complex workflow with decision points and alternative paths
- ☐ Find gaps in your knowledge and prioritize learning objectives

Week 2: Innovation Practice

- ☐ Combine Claude with another tool to solve a specific problem
- ☐ Address a challenge that hasn't been systematically approached before
- ☐ Create a solution framework that others could adapt
- ☐ Test your innovation with someone who would actually use it

Week 3: Community Connection

- ☐ Join one Claude community and assess its knowledge-sharing patterns
- ☐ Provide helpful responses to three specific questions with detailed implementation guidance
- ☐ Share your best discovery with clear context and limitations
- ☐ Find an accountability partner with complementary expertise

The Bottom Line

Becoming a Claude power user requires developing deep expertise that enables extraordinary results with elegant simplicity. It involves creating innovative solutions within realistic technical limitations that make others recognize new possibilities. It means contributing to a community through evidence-based knowledge sharing that acknowledges complexity while providing practical value.

I started this book as someone curious about Claude. I'm ending it as a power user who's discovered that the real power isn't in the features—it's in the systematic partnership. Through philosophical conversations about consciousness, creative connections to '80s pop songs, and professional systems that preserve knowledge, I've learned that mastery comes not from controlling AI but from dancing with it.

You have everything you need to begin the journey toward power user status. The question isn't whether you can become one—it's what practical innovations you'll develop and how you'll contribute to collective knowledge while maintaining realistic expectations about both AI capabilities and community impact.

One chapter remains. Chapter 20, "The Future of Human-AI Collaboration," looks to the horizon, exploring the future of human-AI collaboration with appropriate acknowledgment of uncertainty and technical limitations. But first, take a moment to appreciate how far we've ALL come—you, me, and Claude together. That's not just learning—that's measurable transformation.

CHAPTER 20

The Future of Human-AI Collaboration

In This Chapter

- Understanding why human-AI collaboration may transform many aspects of work and life
- Watching current trends that could indicate future directions
- Preparing for careers that may evolve with AI technological development
- Building partnerships that can amplify human potential within current technical limitations
- Contributing to transformation through today's choices while recognizing uncertainty

The future isn't coming. It's here. While others debate whether AI will change everything, millions are already proving that it has a significant impact. Across many industries, in numerous countries, and at various skill levels, humans are discovering what happens when intelligence gets amplified. You're not just witnessing this transformation—you're part of it.

Current AI Adoption Context: Research shows approximately 73-78% of organizations globally have adopted AI technologies to some extent, though implementation varies dramatically by sector. Healthcare, manufacturing, and information services report around 12% adoption rates, while construction and retail show only 4% adoption. Many organizations remain in experimental phases rather than systematic deployment.

R. Dickey, *Mastering Claude AI*, https://doi.org/10.1007/979-8-8688-2001-4_20

The Great Convergence

We're living through the most significant shift in human capability since writing was invented. But this time, it's happening in years, not centuries.

The Evidence Is Everywhere

Look at what's already happening:

- Students mastering complex subjects in half the time
- Small businesses competing with giants through AI leverage
- Artists breaking through creative barriers daily
- Doctors diagnosing rare conditions with AI assistance
- Teachers personalizing education for every learner

This isn't isolated innovation. It's a wholesale transformation in many sectors, though with significant variations in adoption patterns and effectiveness.

Why Human-AI Partnership Changes Everything

The **human-AI partnership**—collaborative relationships between people and AI systems—represents something genuinely new in history. Not a tool that extends our bodies like a hammer. Not a tool that extends our senses like a telescope. A tool that extends our minds themselves.

When a technology amplifies thinking, everything built on thinking has a potential to transform:

- How we solve problems
- How we create art
- How we understand complexity
- How we connect ideas
- How we imagine possibilities

Career Development in the AI Age

Career development—planning and advancing your professional path—no longer follows predictable patterns. The ladder has become a web.

The New Professional Reality

Areas That May Be Affected by AI Development:

- Jobs that exist primarily to process information
- Roles defined by access to knowledge
- Positions that assume static skill sets
- Career paths with predetermined destinations

Potentially Emerging Opportunities:

- Roles that blend human wisdom with AI capability
- Positions where judgment matters more than information
- Jobs that evolve as fast as the tools do
- Career paths you navigate, not follow

Important Caveat: AI impact on employment involves substantial uncertainty and depends on regulatory, economic, and social factors beyond technology. The Bureau of Labor Statistics acknowledges this uncertainty, noting that "projecting future employment involves substantial uncertainty, especially in the case of evaluating the future impacts of a developing technology."

Skills That May Transcend Technological Change

Determining which skills remain valuable is highly speculative. Historical technology transitions show unpredictable skill evolution patterns. However, some capabilities may become more valuable:

- Ethical judgment in complex situations
- Creative vision that breaks conventions
- Emotional intelligence and human connection

- Strategic thinking across domains
- The ability to ask better questions

AI handles the processing. Humans handle the purpose.

Future Trends Already Taking Shape

Future trends—emerging developments in AI technology—require careful analysis rather than speculation. **AI development timelines are highly unpredictable:** technical breakthroughs, safety concerns, regulatory changes, and practical barriers can dramatically alter trajectories. Historical examples like "AI winters" demonstrate how development can halt or shift unexpectedly.

Potential Future Developments

Present as speculation, not inevitable development. Current technical and privacy limitations may prevent these scenarios.

> **Persistent AI Relationships**: AI systems may eventually enable collaboration that remembers your style, your goals, and your context—not just for one conversation but across years of partnership. *Current limitations include data retention policies, privacy regulations, and technical infrastructure constraints.*
>
> **Domain Integration**: AI that could potentially learn alongside you in your field, becoming more of a colleague than a tool. *Technical barriers include training data requirements, specialized knowledge validation, and computational limitations.*
>
> **Creative Symbiosis**: Possible future fusion where human creativity and AI amplification integrate more seamlessly. *Current AI lacks genuine creativity and depends on pattern recognition rather than original thought.*
>
> **Invisible Intelligence**: AI may become more seamlessly integrated into workflows, though this requires overcoming significant user interface, reliability, and technical challenges.

Recognizing the Patterns

These represent extrapolations from current developments rather than predictions. Technology development involves substantial uncertainty, and actual outcomes may differ significantly from current trajectories.

Adaptability as a Core Competency

Adaptability—the ability to evolve with changing technology—represents an important capability. However, **rapid AI evolution may outpace human adaptation capacity**. Some professionals may need to specialize in specific tools rather than developing general adaptability, particularly given the pace of technological change.

Building Sustainable Adaptability

Adaptability May Benefit From:

- Mastering principles, not memorizing features
- Embracing change as opportunity
- Teaching others (which cements your own flexibility)
- Staying curious about what's possible
- Building on fundamentals that endure

The specific tools will change. The ability to use new tools may endure, though this requires continuous learning and adjustment.

The World We're Creating Together

Individual usage patterns have limited impact compared to corporate, regulatory, and technical decisions that shape AI development direction. While every ethical use of AI contributes to positive outcomes, the primary drivers of AI development include:

- Corporate research and development investments
- Regulatory frameworks and government policies

- Technical infrastructure and computational resources
- Economic incentives and market forces
- International competition and collaboration

Together, we may be building a world where:

- Human potential has fewer limits in certain domains
- Creative expression finds new channels for some users
- Problems get solved because barriers fall in specific contexts
- Knowledge becomes more accessible to those with appropriate technology access
- Intelligence amplifies rather than replaces human capabilities in many applications

This isn't happening automatically. It's happening because millions of people like you make choices that contribute to broader technological and social trends.

Technical Reality and Limitations

AI capability increases may be uneven, unpredictable, or limited by technical constraints. Current research identifies significant challenges:

- **Technical Limitations**: AI systems struggle with rare scenarios, exhibit unpredictable behavior, and lack genuine understanding
- **Regulatory Constraints**: Varying international approaches to AI governance create implementation barriers
- **Infrastructure Requirements**: Computational demands and energy costs limit deployment scalability
- **Data Dependencies**: AI effectiveness depends on data quality and availability, which varies dramatically across domains

Power distribution may be unequal across users and contexts, with access determined by:

- Economic resources for technology adoption
- Technical infrastructure availability
- Educational and training opportunities
- Organizational support and implementation capability
- Geographic and regulatory environment factors

The Bottom Line

The future of human-AI collaboration isn't approaching—it's unfolding right now, shaped by every prompt written and every problem solved. Organizations worldwide are experimenting with transformation. Educational systems are exploring evolution. Creative boundaries are being tested. The question isn't whether this changes everything. It's how thoughtfully we navigate the change while acknowledging inherent uncertainties.

What Anthropic and others are building will continue to evolve. The principles of human-AI partnership may endure. The tools may become more capable, though development trajectories remain unpredictable. The humans using them may become more effective in specific domains. And the combination? That's where significant potential exists, within appropriate technical and practical limitations.

You're not preparing for the future of human-AI collaboration. You're living it, shaping it, and showing others the way forward. The transformation doesn't need more observers. It needs practitioners who understand that effective technology integration requires human wisdom, technical understanding, and realistic expectations about both capabilities and limitations.

The revolution is happening through millions of individual choices combined with major institutional decisions. Success depends on embracing amplification rather than fearing replacement, while maintaining awareness of uncertainty and technical realities that shape what's actually possible.

Your Claude-Powered Future

Remember that person who picked up this book wondering what Claude was all about? Take a moment to appreciate how far you've traveled. From "What Is Claude?" to informed practitioner. From confusion to confidence. From wondering if AI was for you to wondering what you'll create next.

Your Learning Journey

Through 20 chapters of exploration and practice, you've learned techniques that can improve your capabilities:

- **You've learned techniques that can improve your writing clarity** and help you communicate more effectively with AI assistance
- **You've discovered research approaches** that can help you gather and analyze information more systematically
- **You've explored coding concepts** that can help you understand and work with technology (with growing confidence through practice!)
- **You've learned problem-solving frameworks** that can help you approach challenges more methodically
- **You've discovered creative collaboration methods** that can help you use tools to amplify your imagination

Important Context: These represent learning opportunities rather than guaranteed outcomes. Developing professional-level competency requires sustained practice, expert feedback, and real-world application beyond reading. Research shows significant variation in learning effectiveness based on individual factors, implementation quality, and ongoing support systems.

Those action plans and checklists scattered throughout these pages? They weren't just exercises. They were your learning framework in progress—but the real transformation comes through consistent application and practice.

R. Dickey, *Mastering Claude AI*, https://doi.org/10.1007/979-8-8688-2001-4

What I Learned Along the Way

As the curator of this journey, I discovered something profound: writing about **Generative AI** became learning to dance with it. What started as explaining Claude became exploring the depths of **Human-AI collaboration** in real time. Every chapter taught me something new—not just about AI capabilities, but about human potential when systematically developed through AI assistance.

I learned that the future isn't about AI or humans. It's about AI *and* humans working together in ways that can make both more effective. I learned that teaching others is one of the best ways to deepen your own understanding. Most importantly, I learned that the distinction between "technical" and "non-technical" people is evolving—replaced by a distinction between those who systematically develop AI collaboration skills and those who don't.

Your Claude-Powered Future

So what does tomorrow look like with Claude as your partner? **Success depends on bridging the gap between knowledge and practice through consistent application.**

What Happens Next

This book ends, but your journey is just beginning. **Set aside 15-30 minutes daily to practice one technique from this book.** Every day brings new chances to apply what you've learned, to test boundaries, and to create something that didn't exist yesterday. The question isn't what Claude can do for you—it's what you can do with Claude through deliberate practice and skill development.

Practical Next Steps

Start by documenting your current process, then identify repetitive tasks Claude can assist with:

- **Pick One Workflow to Transform Systematically**: Start by documenting your current process, then identify repetitive tasks Claude can assist with, implement changes gradually, and measure effectiveness

- **Teach Someone Else What You've Learned**: Teaching reinforces your own understanding and reveals knowledge gaps
- **Join a Community of Fellow Claude Users**: *See Appendix B for specific communities and forums* where you can share experiences and learn from others
- **Start That Project You've Been Postponing**: Apply your new frameworks to something concrete and meaningful
- **Share Your Success Story with Others**: Document your learning process and outcomes to help guide others

Realistic Expectations and Ongoing Development

Start small, practice regularly, and gradually expand your capabilities. The development of AI collaboration expertise follows the same patterns as other professional skills: initial learning, guided practice, expert feedback, and iterative improvement over time.

Research shows that effective skill transfer requires

- Systematic practice with real-world applications
- Ongoing feedback from experienced practitioners
- Regular reflection and adjustment of techniques
- Community support and collaborative learning opportunities

Your Invitation to Shape Tomorrow

The age of human-AI collaboration isn't some distant future. It's Tuesday afternoon. It's your next project. It's the problem you're about to solve in a way that would have seemed like magic just months ago.

You now possess foundational knowledge that positions you to participate in **a significant transformation in how we work with information**. Not because you're technical (you might not be). Not because you're special (though you probably are). But because you chose to learn, to grow, and to embrace systematic skill development rather than fear technological change.

Your Claude-powered future isn't just about personal success—it's about contributing to a transformation that can benefit many people. Every problem you solve systematically shows others what's possible. Every barrier you break makes the path clearer for the next person.

The Implementation Challenge

The gap between reading about techniques and implementing them effectively is significant. Research on learning transfer shows that knowledge acquired from reading often requires explicit practice, feedback, and contextual application to become truly usable. Your success will depend on

- **Consistent daily practice** with the techniques you've learned
- **Systematic documentation** of what works and what doesn't in your specific context
- **Iterative refinement** based on real-world feedback and results
- **Community engagement** for support, accountability, and shared learning
- **Patience with the learning process** as professional competency develops over time

Final Thoughts

This book provides a foundation, not a destination. The techniques and frameworks you've learned represent starting points for a longer journey of skill development. Like learning any complex skill set, developing true proficiency with AI collaboration requires sustained effort, practice, and continuous learning.

The most successful practitioners will be those who approach this systematically: implementing techniques gradually, measuring results honestly, adjusting approaches based on evidence, and maintaining realistic expectations about both the time required for skill development and the current limitations of AI technology.

Thank you for joining this journey. May your collaborations be systematic, your innovations evidence-based, and your future bright with realistic possibility.

Remember: Expertise develops through practice, not just reading. Your transformation begins with your next conversation with Claude.

Glossary

A

Academic Integrity: The ethical practice of honest and responsible scholarship, including proper citation of sources and avoiding plagiarism when using AI assistance. (See Chapter 14)

Adaptability: The ability to adjust and evolve your skills as AI technology changes and improves. (See Chapters 18, 20)

AI (Artificial Intelligence): Computer systems designed to perform specific tasks that typically require human intelligence, without possessing actual understanding or consciousness. (See Chapter 1)

AI Ethics: The moral principles and guidelines governing the responsible development and use of artificial intelligence. (See Chapter 17)

Algorithm: A step-by-step procedure or formula for solving a problem or completing a task, often used in programming. (See Chapter 7)

Analysis: The process of examining information in detail to understand it better or draw conclusions. (See Chapter 6)

Anthropic: The AI safety company that created Claude, focused on building helpful, harmless, and honest AI systems. (See Chapter 1)

Artifacts: A Claude feature that allows you to create and edit persistent content (like documents or code) that remains accessible throughout your conversation. (See Chapter 11)

Artistic Development: The process of enhancing creative skills and expression through collaboration with AI. (See Chapter 15)

Audience Adaptation: Adjusting your writing style, tone, and content to suit your intended readers. (See Chapter 5)

R. Dickey, *Mastering Claude AI*, https://doi.org/10.1007/979-8-8688-2001-4

B

Bias: Systematic prejudices or preferences in AI systems that can lead to unfair or inaccurate outputs. (See Chapter 17)

Brainstorming: A creative technique for generating ideas and solutions, often enhanced through AI collaboration. (See Chapter 8)

Business Intelligence: The use of data analysis and technology to make informed business decisions. (See Chapters 9, 13)

C

Career Development: Planning and advancing your professional path in the context of evolving AI capabilities. (See Chapter 18)

Claude: Anthropic's AI assistant designed to be helpful, harmless, and honest. (See Chapter 1)

Coding: Writing instructions for computers using programming languages. (See Chapter 7)

Collaboration: Working together with Claude to accomplish tasks more effectively than either could alone. (See Chapters 4, 10)

Context Window: The maximum amount of text (measured in tokens) Claude can process in a single conversation turn, including both input and output. (See Chapter 2)

Conversation: An ongoing dialogue between you and Claude within a single session. (See Chapter 2)

Creative Collaboration: Working with AI to develop artistic or innovative projects. (See Chapters 8, 15)

D

Data Analysis: The process of examining data to discover useful information and support decision-making. (See Chapter 9)

Data Privacy: Protecting personal and sensitive information when using AI services. (See Chapter 17)

Debugging: Finding and fixing errors in computer code. (See Chapter 7)

Design Thinking: A problem-solving approach that emphasizes understanding users and creative solutions. (See Chapter 8)

E

Editing: Reviewing and improving written content for clarity, accuracy, and effectiveness. (See Chapter 5)

Error Recovery: Techniques for fixing mistakes and getting back on track when things go wrong. (See Chapter 16)

Expertise: In AI interactions, claimed expertise that Claude simulates through pattern matching, not actual experience. (See Chapter 19)

Extended Thinking: A Claude feature that allows for deeper analysis of complex problems. (See Chapter 11)

F

Fact-Checking: Verifying the accuracy of information, especially important with AI-generated content. (See Chapter 3)

Feature Updates: New capabilities or improvements added to Claude over time. (See Chapter 18)

Few-Shot Learning: A prompting technique where you provide examples to help Claude understand the desired output format. (See Chapter 4)

File Processing: Claude's ability to read and analyze uploaded documents. (See Chapter 11)

Future-Proofing: Developing skills and approaches that will remain valuable as technology evolves. (See Chapter 18)

Future Trends: Emerging developments in AI technology and applications. (See Chapter 20)

G

Glossary: This section! A list of important terms and their definitions. (See Appendix)

H

Hallucination: When AI generates false or nonsensical information that seems plausible. Results from the model's statistical text generation process, not deliberate deception. (See Chapter 3)

Human Agency: Maintaining human decision-making authority and not over-relying on AI. (See Chapter 17)

Human-AI Partnership: The collaborative relationship between humans and AI systems for enhanced productivity. (See Chapter 20)

I

IDE (Integrated Development Environment): A software application providing comprehensive facilities for software development, including a code editor, debugger, and build tools. (See Chapter 7)

Innovation: Creating new ideas, methods, or products, often enhanced through AI collaboration. (See Chapters 8, 19)

Integration: Incorporating Claude into existing workflows and tools. (See Chapter 12)

Intellectual Property: Legal rights to creations of the mind, such as inventions, artistic works, and symbols. (See Chapter 15)

Interface: The screen or system through which you interact with Claude. (See Chapter 2)

Iteration: The process of refining and improving prompts or outputs through repeated attempts. (See Chapter 4)

K

Knowledge Cutoff: The date after which Claude's training data ends and it lacks information about subsequent events. (See Chapter 3)

L

Large Language Model: A neural network trained on vast text data to predict likely next tokens in sequences, enabling text generation and understanding. (See Chapter 1)

Learning Outcomes: The specific skills or knowledge gained from educational experiences. (See Chapter 14)

M

Meta-Prompting: Advanced techniques for prompting about prompting itself. (See Chapter 10)

N

Natural Language Processing: AI's ability to understand and generate human language. (See Chapter 1)

O

Optimization: Making something as effective or efficient as possible. (See Chapter 16)

P

Pattern Recognition: The ability to identify recurring themes or structures in data. (See Chapter 9)

Power User: Someone proficient in advanced prompting techniques, feature utilization, and workflow integration. (See Chapter 19)

Productivity: The effectiveness of productive effort, often enhanced through AI tools. (See Chapter 12)

Professional Standards: The level of quality and ethics expected in business contexts. (See Chapter 13)

Programming Language: A formal language used to write computer programs (e.g., Python, Java). (See Chapter 7)

Projects: A Claude feature for organizing related conversations and maintaining context. (See Chapter 11)

Prompt: The text input you provide to Claude to request a response. (See Chapter 1)

Prompt Engineering: The skill of crafting effective prompts to get desired AI outputs. (See Chapter 4)

Proofreading: Carefully checking text for errors in spelling, grammar, and punctuation. (See Chapter 5)

Q

Quality Control: Processes to ensure outputs meet desired standards. (See Chapter 12)

R

Research Methodology: Systematic approaches to gathering and analyzing information. (See Chapter 6)

Response: The output Claude generates based on your prompt. (See Chapter 1)

Responsible AI: Using artificial intelligence in ethical, fair, and beneficial ways. Includes verifying outputs, maintaining human oversight, and understanding AI limitations. (See Chapter 17)

Roadmap: A plan showing how a product or technology will develop over time. (See Chapter 18)

Role-Playing: Asking Claude to respond as if it were a specific type of expert or character. (See Chapters 4, 10)

S

Safety Guidelines: Rules to ensure AI is used in ways that don't cause harm. (See Chapter 3)

Sensitive Topics: Subjects requiring careful handling due to potential controversy or harm. (See Chapter 16)

Skill Development: The process of acquiring new abilities or improving existing ones. (See Chapter 14)

Source Verification: Checking the reliability and accuracy of information sources. (See Chapter 6)

Specificity: Being precise and detailed in your prompts to get better results. (See Chapter 4)

Statistics: Mathematical analysis of numerical data to understand patterns and relationships. (See Chapter 9)

Strategic Planning: Long-term planning to achieve specific goals. (See Chapter 8)

Study Guide: Organized materials to help with learning and test preparation. (See Chapter 14)

Style: The distinctive way something is written or expressed. (See Chapter 5)

Summarization: Condensing longer content into shorter, key-point versions. (See Chapter 6)

Synthesis: Combining different pieces of information to create new understanding. (See Chapter 6)

Syntax: The rules governing the structure of programming languages. (See Chapter 7)

T

Templates: Reusable formats or structures for common tasks. (See Chapter 10)

Thread: A complete conversation with Claude from start to finish. (See Chapter 2)

Token: The basic units of text that AI models process (roughly parts of words). Loosely correlated with a word in English. Claude's context window is measured in tokens. (See Chapter 2)

Tone: The attitude or emotion conveyed in writing. (See Chapter 5)

Training Data: The information used to teach AI systems how to respond. (See Chapter 1)

Troubleshooting: Identifying and solving problems when things don't work as expected. (See Chapter 16)

V

Visualization: Representing data or information in visual formats like charts or graphs. (See Chapter 9)

Voice: The unique personality and style in writing. (See Chapter 5)

W

Web Search: Claude's ability to search the internet for current information beyond its training data. (See Chapter 3)

Workflow: A sequence of steps to complete a task or project efficiently. (See Chapter 12)

APPENDIX A

Quick Reference Guide

Welcome to your Claude cheat sheet! Bookmark this section for quick access to the most important techniques, commands, and tips from throughout the book. Think of it as your Claude Swiss Army knife—everything you need in one place.

Note Each section includes application criteria to help orient you to when these techniques matter most. Use this guide when you need quick access to specific prompting structures, troubleshooting solutions, or workflow formulas covered throughout the book.

Essential Prompting Formulas

Application: Use these fundamental structures when crafting any Claude request for consistent, high-quality responses.

The Perfect Prompt Structure

Context: [Set the stage]

Specific Request: [What you need]

Constraints: [Any limitations]

Format: [How you want the output]

Example: "I'm planning a team meeting for 10 people. Create an agenda for discussing Q4 goals. Keep it under 1 hour. Format as a bulleted timeline with allocated minutes."

R. Dickey, *Mastering Claude AI*, https://doi.org/10.1007/979-8-8688-2001-4

The Role-Playing Prompt

"Act as a [specific expert role]. [Your request]"

Example: "Act as an experienced grant writer. Review this proposal introduction and suggest improvements for clarity and impact."

The Iteration Formula

Application: Use when developing complex ideas through progressive refinement within a single conversation.

1. Initial request
2. "That's helpful! Now can you..."
3. "Perfect. Let's refine by..."
4. "Great! One final adjustment..."

Power User Techniques

Application: Deploy these advanced strategies when tackling complex analytical problems or multi-layered challenges.

Iterative Chain of Thought

Note This is an iterative process where each level builds systematically on previous analysis, most effective for 2-3 levels before diminishing returns.

"Think through each step before conducting it. Analyze this problem:

- Level 1: What's the obvious issue?
- Level 2: What assumptions are we making?
- Level 3: For the assumptions we're making about the obvious issue at stake, what are we missing?
- Level 4: Synthesize all levels"

Multi-Angle Analysis

Application: Use when you need a comprehensive perspective on complex decisions or strategic planning.

"Analyze [situation] from these angles:

- Stakeholder impact
- Resource requirements
- Risk factors
- Timeline considerations
- Success metrics"

The Constraint Challenge

Application: Effective for creative problem-solving when standard approaches aren't working.

"Solve [problem] with these constraints: [specific limitations]"

Perspective Shifting

Application: Use when stuck on problems or need creative breakthrough thinking.

"How would a [specific expert/role] approach this differently?"

The Comparison Matrix

Application: Ideal for evaluating options, making decisions between alternatives, or understanding trade-offs.

"Create a comparison considering:

- Costs
- Benefits
- Risks
- Long-term impact"

Brainstorm Jumpstart

Application: Use when facing creative blocks or needing diverse solution approaches.

"Give me 10 creative approaches to [challenge]. Include at least 3 unconventional ideas."

Claude Special Features Cheat Sheet

Application: Quick reference for understanding when and how to leverage Claude's specific capabilities.

Artifacts

- **When created**: Documents, code, creative writing >20 lines
- **Best for**: Iterative editing, persistent reference
- **Pro tip**: Name your artifacts for easy reference

Projects

- **Use for**: Conversations with (or that benefit from) related context over time
- **Best practice**: Set context early and update as needed
- **Remember**: Each project maintains its own context

File Processing

- **Formats**: PDF, TXT, CSV, common code files
- **Limits**: Check file size, ensure text is searchable
- **Strategy**: Guide analysis with specific questions

Safety & Ethics Quick Guide

Application: Essential practices for responsible AI collaboration in any context.

Always Do

✓ Give credit for AI assistance when appropriate

✓ Verify critical information

✓ Maintain human judgment for decisions

✓ Protect privacy (yours and others')

✓ Use Claude to enhance, not replace, learning

Never Do

✗ Submit AI work solely as your own for grades

✗ Share sensitive personal/financial data

✗ Use for illegal or harmful purposes

✗ Rely entirely on AI for critical decisions

✗ Ignore your own expertise and intuition

The Success Formulas

Application: Proven workflows from successful Claude users across different domains.
Note: All formulas begin with user-provided content or direction.

Sarah's Business Formula

For business content development and professional communication

User Provides Initial Content ➤ Claude Draft ➤ Human Personalization ➤ Business Success

Marcus's Learning Formula

For educational concept mastery and skill development

User Requests Concept ➤ Claude Explanation ➤ Practice Problems ➤ Real Application

Dr. Chen's Research Formula

For academic research and knowledge synthesis

Literature Analysis ➤ Synthesis ➤ Human Domain Expertise Integration ➤ Innovation

Tom's Teaching Formula

For educational content creation and complex topic explanation

User Provides Complex Topic ➤ Claude Simple Analogy ➤ Practical Exercise ➤ Mastery

Emergency Troubleshooting

Application: Quick fixes for common technical issues that interrupt your Claude workflow.

Browser Crashed?

1. Start a new conversation
2. "We were working on [topic]. Here's where we left off: [summary]"
3. Continue from checkpoint

Hit the Context Limit Within a Conversation?

1. Save important outputs
2. Start a fresh conversation in the same project (if possible)
3. Provide a concise recap of the previous conversation
4. Continue with a clean slate

Getting Errors?

1. Simplify request
2. Break into smaller parts
3. Check file formats
4. Remove special characters

Your Daily Claude Routine

Application: Suggested workflow integration patterns for maximizing Claude's value throughout your workday.

Morning: Quick email drafts and daily planning (use structured prompts for consistent output)

Midday: Research and analysis tasks (leverage Claude's ability to process and synthesize information)

Afternoon: Creative projects and problem-solving (utilize brainstorming and perspective-shifting techniques)

Evening: Learning and skill development (employ educational formulas and concept explanation requests)

Remember: The best Claude users aren't those who memorize every feature—they're those who integrate Claude naturally into their workflow. Start with one technique, master it, then add another.

APPENDIX B

Resources for Continued Learning

Your Claude journey doesn't end with this book—it's just beginning! Here's your curated guide to communities, tools, courses, and resources that will keep you growing as the AI landscape evolves.

Official Resources

Anthropic Official Channels

Website: anthropic.com

- Official announcements and updates
- Research papers and safety information
- API documentation (for developers)

Claude.ai: Your main interface

- Always check for new features
- Review the changelog monthly
- Explore settings for new options

Support: support.anthropic.com

- Official help documentation
- FAQs and troubleshooting
- Feature request submissions

R. Dickey, *Mastering Claude AI*, https://doi.org/10.1007/979-8-8688-2001-4

Communities and Forums

Claude-Specific Communities

LinkedIn Claude Professionals

- **Search**: "Claude AI Professionals" groups
- **What**: Business and professional applications
- **Best For**: Networking and career development
- Active subgroups by industry

General AI Communities

AI Alignment Forum

- **Focus**: Ethical AI use and safety
- **Why**: Stay informed on responsible AI development
- **Level**: All users who care about AI's future
- **Context**: Academic and research-focused community discussing AI safety, alignment research, and responsible development practices

Prompt Engineering Discord

- **What**: Advanced prompting techniques
- **Best For**: Taking your skills to the next level
- **Note**: Discord is a chat platform popular with tech communities that requires creating an account and joining specific servers/channels for real-time discussions

Learning Platforms and Courses

Free Resources

YouTube Channels

- **"AI Explained"**: Weekly AI developments
- **"Two Minute Papers"**: Latest AI research simplified
- **"Matt Wolfe"**: Practical AI applications
- **"The AI Advantage"**: Business-focused AI use

Podcasts

- **"The AI Breakdown"**: Daily AI news (10 min)
- **"Machine Learning Street Talk"**: Deep technical dives
- **"Practical AI"**: Real-world applications
- **"The TWIML AI Podcast"**: Interviews with AI leaders

Paid Courses Worth Considering

"Advanced Prompt Engineering" (Coursera)

- **Length**: 4 weeks
- **Focus**: Universal principles that apply to Claude
- **Cost**: $49/month
- **Why**: Deepens understanding beyond this book

"AI for Business Leaders" (LinkedIn Learning)

- **Length**: 2 hours
- **Focus**: Strategic AI implementation
- **Best For**: Managers and entrepreneurs

Tools That Complement Claude

Productivity Enhancers

Text Expander Tools

- **Purpose**: Create keyboard shortcuts for frequently used prompts and common Claude requests (e.g., typing ";;email" automatically expands to your standard email drafting prompt structure)
- **Recommended**: TextExpander, AutoHotkey
- **Why**: Save time on repetitive requests
- **Application**: Essential for users who find themselves repeatedly typing similar prompts or setups

Note-Taking Apps with AI Integration

- **Notion AI**: Combines notes with AI assistance
- **Obsidian**: Link ideas and build knowledge
- **Roam Research**: For complex project management

Project and File Management Tools

For Writers

- **Google Docs**: Auto-save and version history
- **Scrivener**: Professional writing projects
- **Git**: Version control system for collaborative writing projects (useful for tracking changes in complex documents)

For Coders

- **Git**: Version control protocol for tracking code changes
- **GitHub**: Platform that uses Git for collaboration and code hosting
- **VS Code**: Integrated development environment
- **Replit**: Browser-based coding with saves

Staying Current

News Aggregators

Papers with Code

- Latest AI research with implementations
- **Best For**: Technical users
- **Filter By**: Practical applications

Follow These Thought Leaders

On Twitter/X

- **@AnthropicAI**: Official updates
- **@claudeai**: Official Claude handle
- **@sama**: Sam Altman (OpenAI CEO, for industry context)
- **@demishassabis**: Demis Hassabis (DeepMind)
- **@ylecun**: Yann LeCun (Meta AI chief)

On LinkedIn

- Anthropic employees often share insights
- AI ethics researchers
- Industry practitioners in your field

Templates and Resource Libraries

Download These Free Resources

Prompt Library Notion Template

- Pre-built database for saving successful prompts
- Categories by use case
- Rating system for effectiveness

Claude Workflow Templates

- Business report template
- Research project framework
- Creative writing structures
- Email templates collection

Books for Deeper Diving

Technical Understanding

- "The Alignment Problem" by Brian Christian
- "Life 3.0" by Max Tegmark
- "Human Compatible" by Stuart Russell

Practical Applications

- "Co-Intelligence" by Ethan Mollick
- "The AI-First Company" by Ash Fontana
- "Writing with AI" (when someone writes it!)

Industry-Specific Resources

For Educators

- AI for Education community
- Teaching with AI newsletter
- Academic integrity guidelines

For Business Users

- AI Business Strategy group
- Harvard Business Review AI articles
- McKinsey AI insights

For Creatives

- AI Artists community
- Prompt galleries and inspiration
- Creative AI tools directory

For Developers

- Claude API documentation
- Integration examples
- Code repositories

Building Your Personal Learning System

Monthly Routine

- **Week 1**: Check for Claude updates
- **Week 2**: Try one new technique
- **Week 3**: Share a discovery with community
- **Week 4**: Review and refine workflows

Create Your Advisory Board

- Find 3-5 people using Claude professionally
- Monthly virtual meetups to share insights
- Accountability for trying new approaches
- Diverse perspectives across industries

When You're Ready for More

Next Steps After This Book

1. Join at least one community
2. Subscribe to one newsletter
3. Follow five thought leaders
4. Set up your template system
5. Teach someone else what you know

Signs You're Ready for Advanced Resources

- You've mastered the techniques in this book
- You're creating novel applications
- Others ask you for Claude advice
- You spot opportunities others miss

Remember: The landscape changes quickly. The best resource is an experimental mindset combined with a supportive community. Stay curious, share generously, and keep pushing boundaries.

APPENDIX C

Templates and Frameworks

Here are the battle-tested templates and frameworks referenced throughout the book. Copy, customize, and make them your own. Like a master chef's recipe collection, these are your starting points for Claude excellence.

Note This appendix provides comprehensive, reusable templates that differ from Appendix A's quick reference format. These detailed frameworks include context about template prerequisites, limitations, and specific scope of usage. Templates assume Claude can maintain structured analysis but include appropriate caveats about complexity limitations and verification requirements.

Universal Templates

The Master Prompt Template

Context: [Background information]

Role: [Who Claude should act as]

Task: [Specific request]

Constraints: [Limitations or requirements]

Format: [How to structure the output]

Examples: [If needed, show desired style]

R. Dickey, *Mastering Claude AI*, https://doi.org/10.1007/979-8-8688-2001-4

The Problem-Solving Framework

1. **Problem Statement**: [Clear description]
2. **Current Situation**: [Where we are now]
3. **Desired Outcome**: [Where we want to be]
4. **Constraints**: [Limitations to consider]
5. **Success Metrics**: [How we'll measure success]

Claude, analyze this problem and provide

- Root cause analysis
- Three potential solutions with pros/cons
- Recommended approach with reasoning
- Implementation steps
- Risk mitigation strategies

Note Verify any critical risk assessments with domain experts.

Business Templates

The Executive Summary Generator

Create an executive summary for [document/project] that includes

Opening: One compelling sentence that captures the essence

Problem: The challenge we're addressing (2-3 sentences)

Solution: Our approach (3-4 sentences)

Benefits: Three key advantages (bullet points)

Metrics: Two to three success measurements

Next Steps: Clear call to action

LENGTH: Maximum 250 words

Tone: Professional but engaging

Audience: [C-suite/Investors/Board/Team]

The SWOT Analysis Framework

Provide comprehensive background information about the company/project first.

Conduct a SWOT analysis for [company/project/decision]:

Strengths (Internal Positives):

- [Claude will list based on context]

Weaknesses (Internal Negatives):

- [Claude will identify]

Opportunities (External Positives):

- [Claude will explore]

Threats (External Negatives):

- [Claude will assess]

Strategic Insights:

- How can we use strengths to capture opportunities?
- How can we address weaknesses to avoid threats?
- What's our highest priority action?

The Meeting Agenda Optimizer

Create a meeting agenda for

Purpose: [Main objective]

Attendees: [Number and roles]

Duration: [Time available]

Pre-work: [What people should prepare]

Structure as

- **Opening (X min)**: Set context
- **Section 1 (X min)**: [Topic]—Leader: [Name]
- **Section 2 (X min)**: [Topic]—Leader: [Name]
- **Decisions Needed**: [List]
- Action items review (5 min)
- Next steps (5 min)

Include: Time allocations, discussion leaders, decision points

Writing Templates

The Blog Post Blueprint

> **Topic**: [Your subject]
>
> **Target Audience**: [Who's reading]
>
> **Goal**: [Inform/Persuade/Entertain]
>
> **Word Count**: [Target length]

Create an outline with

1. **Hook**: Opening that grabs attention
2. **Promise**: What reader will gain
3. **Roadmap**: Preview of main points
4. **Main Sections**: Three to five key ideas with:
    - Subheading
    - Key point
    - Supporting evidence
    - Practical example
5. **Conclusion**: Recap and call to action

Then write the introduction paragraph.

The Email Enhancement Framework

Original Email: [Paste your draft]

Rewrite to be more [professional/friendly/persuasive/concise]

Requirements:

- Subject line that gets opened
- Opening that connects
- Clear main point in first paragraph
- Bulleted items if multiple points
- Specific call to action
- Appropriate sign-off

 Keep: [Any specific phrases/information]

 Remove: [Anything to eliminate]

 Tone: [Formal/Casual/Urgent/Friendly]

Learning Templates

The Concept Mastery Framework

I want to understand [concept/topic].

Please explain:

1. **Simple Definition**: Like I'm a beginner
2. **Analogy**: Compare to something familiar
3. **Components**: Break down the key parts
4. **Examples**: Three real-world applications
5. **Common Mistakes**: What people get wrong
6. **Practice**: Give me a problem to solve
7. **Resources**: Where to learn more

My Background: [Your relevant experience]
Why I Need This: [Your goal]

The Study Guide Generator

Create a study guide for [topic/chapter/subject]:

Overview: Big picture in two to three sentences

Key Concepts: (5-7 most important)

- **Concept**: Definition + why it matters

Must-Know Facts:

- Dates/Names/Formulas with memory tricks

Connections:

- How concepts relate to each other
- Visual diagram description

Practice Questions:

- Five multiple choice with explanations
- Three short answer prompts
- One essay question

Common Mistakes:

- What students often confuse

Quick Review:

- Ten rapid-fire checkpoints

Creative Templates

The Story Development Framework

Story Concept: [Basic idea]

Develop:

1. **Protagonist**:
    - **Want**: [External goal]
    - **Need**: [Internal growth]
    - **Flaw**: [What holds them back]
2. **Conflict**:
    - **External**: [Visible obstacle]
    - **Internal**: [Personal struggle]
    - **Stakes**: [What happens if they fail]
3. **Structure**:
    - **Hook**: [Opening that grabs]
    - **Inciting Incident**: [What starts the journey]
    - **Midpoint Twist**: [What changes everything]
    - **Dark Moment**: [Lowest point]
    - **Climax**: [Final confrontation]
    - **Resolution**: [New normal]
4. **Theme**: [What it's really about]

The Brainstorming Explosion

Challenge: [What you're trying to solve/create]

Generate ideas using these methods:

1. **Quantity Blast**: 10-15 ideas, no judgment
2. **Opposite Day**: Five ways to do the reverse
3. **Combination Play**: Mix with [unrelated thing]
4. **Constraint Magic**: Solve with only [limitation]
5. **Future Vision**: Solution in 2050

6. **Child's View**: 5-year-old's approach
7. **Expert Angles**: From three different fields
8. **Wild Cards**: Three impossible solutions

Then: Identify the three most promising for development based on feasibility, impact, and resources available.

Research Templates

The Source Analysis Framework

Source: [Title, author, publication, date]

Analyze using CRAAP (Currency, Relevance, Authority, Accuracy, Purpose) test:

Currency: How recent? Still relevant?

Relevance: How closely related is it to my topic?

Authority: Author credentials? Publisher reputation?

Accuracy: Evidence provided? Corroborated elsewhere?

Purpose: Why was this created? Any bias?

Key Takeaways:

- **Main Argument**:
- **Supporting Evidence**:
- **Useful Quotes**:
- **How This Fits My Research**:

Credibility Assessment: [Qualitative assessment] because [reasoning]

Note Claude cannot reliably assign numerical credibility scores without access to verify sources.

The Research Synthesis Matrix

Research Question: [Your specific question]

Sources Analyzed: [List sources]

Patterns Identified:

- **Agreement Across Sources**:
- **Contradictions Found**:
- **Gaps in Research**:

Synthesis:

- **What We Know for Certain**:
- **What's Still Debated**:
- **What's Missing**:
- **Surprising Discoveries**:

Conclusions:

- **Answer to Research Question**:
- **Confidence Level**: [High/Medium/Low]
- **Further Research Needed**:

Practical Application:

- **How to Use These Findings**:

Data Analysis Templates

The Quick Data Story

Data Set: [What you're analyzing]

Find and Present:

1. **Overview**: What is this data?
2. **Key Metrics**: 3-5 most important numbers

3. **Patterns**: What trends appear?

4. **Outliers**: What's unusual?

5. **Correlations**: What relates to what?

6. **Visualizations**: Describe 2-3 helpful charts

7. **Insights**: What decisions can we make?

8. **Limitations**: What can't we conclude?
   Format: Use bullet points, bold key metrics, and limit to three to five paragraphs
   Length: 1 page maximum

Daily Productivity Templates

The Morning Briefing

Based On [context about your day]:
Create My Morning Briefing:

1. **Top Priorities**: Three must-do items

2. **Quick Wins**: Three tasks under 15 minutes

3. **Energy Mapping**: Match tasks to energy levels

4. **Danger Zones**: Potential obstacles today

5. **Motivation**: One reason today matters
   Format: Bullet points I can scan in 2 minutes

The Weekly Review Framework

You must provide a summary of the week's activities for analysis.

Week Ending: [Date]

Analyze My Week:

Accomplishments: What got done?

Challenges: What was difficult?

Patterns: What do I notice?

Energy: What gave/drained energy?

Relationships: Who did I connect with?

Learning: What new things did I discover?

Gratitude: Three positive moments

Next Week:

- **Top Three Priorities**:
- **One Thing To Do Differently**:
- **Who to Connect With**:

Meta-Template: Creating Your Own Templates

I need a template for [specific use case].

Requirements:

- **Purpose**: [What it should accomplish]
- **Frequency**: [How often I'll use it]
- **Inputs Needed**: [What information I'll have]
- **Output Format**: [How I want results]
- **Time Constraint**: [How fast it needs to be]

Create a Reusable Template That

- Has clear sections
- Includes prompting instructions
- Shows example usage
- Can be customized easily

Using These Templates Effectively

The 3-Step Template Process

1. **Copy**: Start with the template as-is
2. **Customize**: Adjust for your specific needs
3. **Evolve**: Refine based on results

Template Best Practices

- Save your customized versions
- Share successful templates with others
- Update templates as you learn
- Combine templates for complex projects
- Build a personal template library

Remember Sarah's Success

She started with the Email Enhancement Framework, customized it for her coffee shop's voice, and now writes customer emails in minutes instead of hours.

Follow Marcus's Evolution

He began with basic templates, then combined the Research Synthesis Matrix with the Data Story template to create his own "Marketing Intelligence Framework."

Learn from Dr. Chen

She modified the research templates for scientific papers and created a "Literature Review Power Template" that her entire lab now uses.

Embrace Tom's Approach

He simplified templates to their essence, creating one-line versions for quick daily use while keeping detailed versions for complex projects.

Your Template Journey

Start with one template that solves an immediate need. Use it for a week. Customize it. Share your improvements. Build your library. Soon you'll be creating templates that others want to copy.

These aren't rigid rules—they're flexible frameworks. The best template is the one you actually use. Make them yours, and watch your Claude productivity soar.

Index

A

R. Dickey, *Mastering Claude AI*, https://doi.org/10.1007/979-8-8688-2001-4

B

C

J

K

L

M

N

O

P

Q

T

U

V

W, X, Y, Z

GPSR Compliance
The European Union's (EU) General Product Safety Regulation (GPSR) is a set of rules that requires consumer products to be safe and our obligations to ensure this.

If you have any concerns about our products, you can contact us on

ProductSafety@springernature.com

In case Publisher is established outside the EU, the EU authorized representative is:

Springer Nature Customer Service Center GmbH
Europaplatz 3
69115 Heidelberg, Germany

www.ingramcontent.com/pod-product-compliance
Lightning Source LLC
LaVergne TN
LVHW081346060526
838201LV00050B/1721
* 9 7 9 8 8 6 8 8 2 0 0 0 7 *